Survival by Hunting

Survival by Hunting

*Prehistoric Human Predators
and Animal Prey*

George C. Frison

UNIVERSITY OF CALIFORNIA PRESS
Berkeley • *Los Angeles* • *London*

University of California Press
Berkeley and Los Angeles, California

University of California Press, Ltd.
London, England

© 2004 by the Regents of the University of California

Library of Congress Cataloging-in-Publication Data
Frison, George C.
    Survival by hunting: prehistoric human predators
and animal prey / George C. Frison.
        p.   cm.
    Includes bibliographical references and index.
    ISBN 0-520-23190-2 (cloth : alk. paper).
    1. Hunting, Prehistoric.   2. Paleo-Indians—
Hunting.   3. Hunting and gathering societies.
I. Title.
GN799.H84F75   2004
306.3'64—dc22   2003018996

Manufactured in the United States of America

13   12   11   10   09   08   07   06   05   04
10 9 8 7 6 5 4 3 2 1

To June,
my spouse for more than half a century,
and my paternal grandparents,
Jacob and Margaret Frison

Contents

Illustrations and Table

Preface

My awareness of the need for a book on prehistoric human hunting has evolved gradually over more than three decades of teaching and research. This awareness is the result of dissatisfaction with ethnographic and archaeological interpretations of human hunters and hunting that fail to acknowledge the years of experience and the accumulation of knowledge of animal behavior required to become a successful hunter. Particularly annoying is the reliance on artists' unrealistic portrayals of prehistoric hunting episodes that become imprinted on the minds of viewers, especially students, and are almost impossible to eradicate.

Equally disturbing to me is the attitude students are acquiring toward hunting. Faunal studies have steadily increased the information that can be derived from bone beds in archaeological kill sites, and during the past three decades, taphonomy has become a substantial part of the jargon of archaeology. Faunal analysis, especially in communal kill sites containing the remains of large numbers of animals, can reveal many past activities, including the seasonality of killing events, animal population structure, butchering and processing techniques, and the amount of nutrients provided to the hunters. Yet faunal analysis tells us little about what happens from the time the hunter or hunters leave their camp until the hunt is terminated. Thus students questioned about animal procurement strategies commonly respond, "When they got hungry, someone would simply kill a bison or whatever other animal was selected as the target for the day and bring it back to camp." I believe such interpretations to be totally inadequate, and I hope that the contents of this book convince others of the vast reservoir of learned behavior involved in hunting.

Those contents are derived mainly from two sources. One is the archaeological record of large mammal hunting in a part of western North America; the other is my personal experiences, which began at a time and place in which hunting was an integral part of the economic process. The time was the early 1930s and the place was in northern Wyoming, where the combined effects of severe drought and economic depression resulted in an increased dependence on wild animals for food; at the same time, the survival of domestic animals on the open range required their constant protection from predators.

As I recall my earliest experiences in academic archaeology, most published interpretations of prehistoric hunting strategies appeared to me inadequate and often inaccurate. The data coming out of the ground revealed much of what happened but not how it happened. The participant ethnographic analogies of Richard Lee and Irven DeVore (1968) and Lewis Binford (1978) seemed to me a step in a positive direction, but the former dealt with African Bushmen and the latter with Eskimo caribou hunters. Though both contained observations relevant to hunting in general, I found them difficult to apply to Great Plains and Rocky Mountain prehistoric hunters.

The multidisciplinary approach was another positive research strategy, intended to reconstruct the past ecosystems within which prehistoric human groups lived. Geologists reconstructed old landforms; palynologists reestablished old plant communities; and biologists used present-day small mammal ecology to pinpoint similar conditions in the past. Borrowing heavily from paleontology, taphonomists began to examine archaeological bone beds and develop innovative methods of identifying the age of animals to determine the seasonality of procurement activities, establish animal population profiles, and distinguish between human, animal, and natural modification of faunal remains.

Wildlife biologists interested in behavioral studies were yet another source of valuable information. Their efforts brought archaeologists another step closer to the reality of prehistoric animal procurement; but unless a particular wildlife biologist's interests happened to lie in the area of human predator–large animal prey relationships (i.e., hunting), he or she could supply very few details on prehistoric hunting strategies. After reviewing the literature, I became convinced that an experienced modern hunter could raise practical questions and provide possible answers. An intimate knowledge of animal habitats and behavior, expertise in the use of weaponry and tools, and experience acquired through both hunting and predator control might lead to a better understanding of past hunting strategies. These are the topics that shape the following discussion.

Acknowledgments

Numerous individuals have influenced my thoughts concerning pre-historic hunting. My grandfather, Jacob Frison, began my enthusiasm for wild animals and hunting at an early age. He, along with an uncle, Theodore (Ted) Frison, guided me through my early hunting years. In the past few decades, George Zeimens of Torrington, Wyoming, helped me realize the advantages gained when two closely cooperating hunters, both familiar with animals and their territory, pool their efforts.

William Mulloy, University of Wyoming; Preston Holder, University of Nebraska; and H. Marie Wormington, Denver Museum of Natural History, provided the encouragement for someone then considered a non-traditional student to embark on an academic career. James B. Griffin and Arthur Jelinek at the University of Michigan continued that en-couragement through graduate school. After I finished graduate school, Paul McGrew, a paleontologist at the University of Wyoming, and I—along with Charles Reher, Danny Walker, and Michael Wilson, who were among my first graduate students at the University of Wyoming—were able to initiate methods of faunal analysis that, although improved on since then, are still being used to good advantage. Working with C. Vance Haynes, Jr., of the University of Arizona, and John Albanese, a consult-ing geologist in Casper, Wyoming, I rapidly perceived the importance of geological expertise in the identification of landforms used by humans in animal kills.

I am indebted to the Smithsonian Institution for research support and to Dennis Stanford, Waldo Wedel, and Clifford Evans, all from the Smithsonian, for access to Paleoindian collections housed there. Throughout the years, research support came from the National Science Foundation, the National Geographic Society, the L. S. B. Leakey Foundation, the U.S. National Park Service, the National Endowment for the Humanities, the U.S. Bureau of Land Management, the U.S. Bureau of Reclamation, the U.S. Fish and Wildlife Service, the University of Wyoming, the Wyoming Archaeological Foundation, the Wyoming Archaeological Society, the Wyoming Council for the Arts, the Wyoming Council for the Humanities, the Wyoming Recreation Commission, and the Colorado Historical Society. Private research support came from varied sources, including the Carter Mining Corporation, the Kerr-McGee Corporation, Joseph Cramer, Forrest Fenn, Mark Mullen, Mike Kammerer, Jack Krmpotich, and William Tyrrell.

The Wyoming Game and Fish Department has continually cooperated in providing faunal specimens for what has become a major large mammal comparative faunal collection at the University of Wyoming. Ned Frost and Roy Coleman of Cody, Wyoming, were guides and outfitters who acquainted me with a large part of the high country in northwest Wyoming and shared their experiences with mountain sheep and grizzly bear. Kay Bowles was the manager of the mountain sheep herd on Whiskey Mountain at Dubois, Wyoming, and allowed me to participate in the trapping of winter sheep to transplant them elsewhere. He and Amos Welty, also of Dubois, took me on several trips into the high country to visit sheep traps. Over a period of several years, Pete (Bison Pete) Gardner of Wheatland, Wyoming; Arthur Buskohl of Gillette, Wyoming; and George Crouse of Laramie, Wyoming, acquainted me with the ways of bison. Dale Guthrie of the University of Alaska gave me a short introduction to Dall sheep hunting in Alaska.

Olga Soffer, University of Illinois, and Nikolai Praslof, Russian Academy of Sciences, arranged access to preserved mammoth remains in the zoological collections in St. Petersburg. Gary Haynes, University of Nevada, Reno, smoothed the way for me to participate in elephant culls in Zimbabwe in 1984 and 1985. Besides introducing me to African wildlife, Clem Coetzee, a wildlife manager in Hwange National Park, Zimbabwe, generously provided food and lodging while we were on the elephant culls. Bruce Bradley, an archaeological consultant in Cortez, Colorado, provided the fluted points used in my weaponry experiments. Robert Cole, of Thermopolis, Wyoming, aroused my early interest in the manufacture and use of the bow and arrow.

Anne Slater, University of Wyoming, and William Woodcock, a consulting editor in Berkeley, California, spent many hours reading, correcting bad grammar, and pointing out inconsistencies in my thinking. Lee Lyman, University of Missouri, and an anonymous reviewer made many valuable comments on the first draft of the manuscript. Line drawings of stone projectile points were made by Connie Robinson, a very talented commercial artist from Sheridan, Wyoming.

Down through the years, I have had the benefit of excellent students and volunteers on field crews. I acknowledge all the persons who allowed access to their land, permitted the use of buildings and equipment, and in innumerable ways extended their services toward my involvement in archaeology. I thank my wife of more than a half century, June Frison, for sharing and thus tolerating my passion for both hunting and archaeology, which has required long periods away from home. To these and many others, my sincere thanks. Their help, thoughts, and ideas have influenced my thinking, but the contents of this book reflect how my own hunting experiences and involvement with animals have affected my interpretation of the archaeological record as it pertains to prehistoric hunters and their subsistence strategies.

CHAPTER I

Where the Buffalo
Once Roamed

THE HISTORICAL BACKGROUND

My paternal grandparents, Jake and Margaret Frison, were true pioneers. Jake was a railroad engineer, and he and my grandmother married and moved to Leadville, Colorado, in 1890. Their dream of the future was to own a cattle ranch; so three years later, they purchased a small place along the Roaring Fork River at Basalt, Colorado. However, they were not able to expand the property into the kind of ranch operation they were searching for, and in 1901, in their late thirties and with four young children, they decided to abandon a secure but to them unsatisfying life in Colorado to move to northern Wyoming. After spending most of the summer of 1900 traveling through several western states, and after considering several possible locations, they finally chose a spot in the Big Horn Basin in north-central Wyoming. In the summer of 1901, they gathered up their small herd of cattle, loaded their belongings on wagons, and began the trek from Basalt to Ten Sleep, a distance of about 500 kilometers as the crow flies but considerably further on the long, winding roads of that time. Following a route that took most of the summer of 1901, they reached their destination; the home ranch was to be at the base of the western slope of the Big Horn Mountains along a flowing mountain stream with mountains to the east and plains to the west. Even while facing a long, cold Wyoming winter, they were making plans to begin acquiring the range land needed to support cattle ranching.

By 1918, the ranch was a modest but viable cattle operation. Also by this time, each of the couple's three sons, one of them my father, had completed all necessary requirements for owning a homestead of 640 acres, which together formed a block of nearly 2,000 acres of prime mountain range. It was a transhumance operation: cattle were taken to the mountains in summer and to the plains in winter. Even relying only on horses for transportation, one could gather cattle from the plains one day and move them to the mountains the next. Besides the domestic animals, deer, elk, and pronghorn, along with a host of predators, occupied the area. My grandfather was an avid hunter and trapper: I am sure the presence of wild game strongly influenced his final selection of a place to put down permanent roots. The family's way of life was harsh but still rewarding because of the unfettered access and freedom found in the wide open spaces.

My father was killed in an accident in 1924, just before I was born, and my mother could not visualize a future for herself on the ranch. Consequently, when I was three, my mother left and my grandparents took on the chore of raising me. I took to ranch life like a duck to water from the day I was placed on the back of a horse and followed my grandfather around the ranch. As soon as I could put a saddle and bridle on a horse by myself, a whole new world opened up. At an early age, I began to see the plains and the mountains not as two separate ecosystems but as a continuum, a concept I have used repeatedly in analyzing plains and mountain prehistory. To me, the two ecosystems were inseparable, and ranchers had to deal with them as complements; the same was undoubtedly true for prehistoric hunters. Other ecologists have arrived at similar conclusions (see, e.g., Knight 1994).

THE PLAINS AND THE MOUNTAINS

The late Waldo Wedel, the first widely recognized authority on Great Plains archaeology, referred to the area as a land of sun, wind, and grass (Wedel 1961). These are appropriate terms, but they strike me as inadequate to convey a realistic picture. Just a few of the obvious characteristics missing from his description are bitterly cold winters; oppressively hot summers, with hailstorms and tornadoes; spring, fall, and winter blizzards; abrupt weather changes; wood ticks; swarms of biting insects; and rattlesnakes. It is a land of grass but there are also seemingly endless stretches of sagebrush, yucca, greasewood, salt bushes, and juniper, along with many areas of soil incapable of producing any vegetation. Livestock raisers and wildlife managers have as many ways of describing grass as the Inuit have to describe snow—conditions that, although very different, are critical to the daily lives of both.

One has to be close to the land at all times of the year, over many years, to acquire a true feeling for it and be able to extract a living from it. For all their good and bad qualities, the plains and the Rocky Mountains together provide a large share of the information about prehistoric large mammal hunting during the more than 11,000 years of known human habitation in North America. This is where the evidence for the hunting of these animals is found in good geologic contexts. Yet the records of the livestock industry over more than a century suggest that prehistoric human survival must have been tested on many occasions.

In reality, the plains and Rocky Mountains are composed of an almost endless variety of landforms—flat to rolling plains, mountain ranges, isolated uplifts, semideserts, playa lakes, vegetated sand hills, active sand dunes, foothills, intermontane basins, flowing springs, intermittent streams, permanent watercourses, large rivers, dry arroyos, swamps, mountain meadows, buttes, deep canyons, glacial features, high peaks, and year-round snow fields. However, to avoid endless description of landscapes that are peripheral to the main topic of prehistoric hunting, I will reduce the area of reference to the commonly recognized physiographic regions of the Great Plains, Rocky Mountains, Great Basin, Colorado Plateau, and Columbia Plateau. Within these are several small geographic locations that I believe have yielded information pertinent to prehistoric hunting; these include the Wyoming Basin and the Big Horn Basin, both in Wyoming; the Black Hills, situated mostly in South Dakota and partly in Wyoming, with a small extension into Montana; Middle Park and the San Luis Valley, both in Colorado; and the Yellowstone Plateau in Wyoming, Idaho, and Montana (map 1). Continuing research in plains and mountain archaeology will undoubtedly result in the recognition of other pertinent geographic areas.

The wide variety of physiographic features provides very different and rapidly changing vegetative cover critical to animal ecology. The wide expanses of open plains are mostly treeless and, depending on soil conditions and moisture, are covered with several different species of grasses, yucca, and sagebrush. Low-lying and poorly drained areas support alkali-tolerant greasewood and salt bushes. In contrast, riparian areas along river valleys with terrace systems support trees, shrubs, and tall grasses. Foothills are carved by arroyos of varying sizes and depths, depending on gradients and underlying bedrock: foothill vegetation consists of

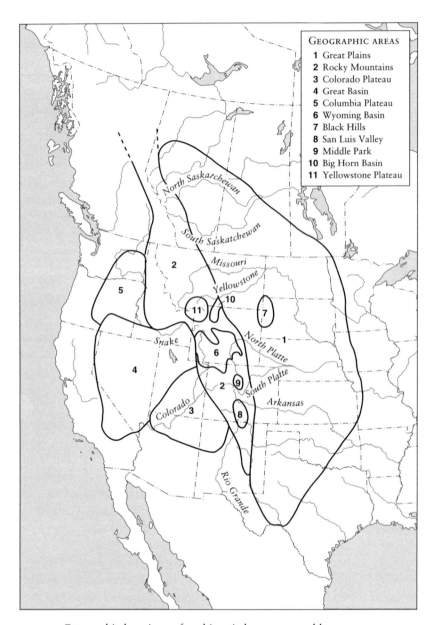

MAP 1. Geographic locations of prehistoric large mammal hunters.

juniper, mountain mahogany, scattered trees, and bunch grasses. There is an arbitrary and not always well-defined boundary between upper foothills and mountain meadows and slopes; in places, it extends to the timberline. Increased moisture at the higher elevations produces pine and aspen forests interspersed with open parks that support tall, dense stands of grass and big sagebrush. Alpine vegetation begins at the timberline and continues up to more than 4,000 meters (13,000 ft.).

Thick stands of trees and browse may be found on north- and east-facing mountain slopes, because there less moisture is lost through evaporation. On south- and west-facing slopes, drier conditions result in scattered pines, junipers, mountain mahogany, sagebrush, and bunch grass. Certain sedimentary sandstone and limestone formations are aquifers; in many locations, where exposed, flowing springs provide water for all animals. Mountain ranges form rain shadows that deny rainfall to certain areas; for example, parts of the interior of the Big Horn Basin in northern Wyoming, aptly referred to as badlands, receive as little as 12 centimeters of annual precipitation and are more reminiscent of the Great Basin than the Great Plains. Mountain ranges also affect the movement of storm systems, influencing both winter snow accumulations in higher elevations and rainfall patterns during the warm months. The open plains and foothills usually become dry by early summer, and vegetation turns brown. Meltwater from deep winter snow in the mountains feeds and flushes out the river systems and provides moisture for trees, brush, and tall grasses on banks and floodplains; these may remain green throughout the summer, in marked contrast to areas that lack adequate moisture (see Knight 1994).

Seasons are clearly defined over the entire area, and one must frequent it at all times to gain a basis for understanding the significance of these changes to the hunters who lived there in the past. There is a stark contrast between the long, hot days of summer and the extended periods of sub-zero temperatures in winter. Everyone hopes that warm, sunny days in early spring and late fall presage early grass and mild winters, respectively; but more often than not, these short periods of good weather rapidly deteriorate into blizzard conditions hazardous to animals both wild and domestic. Having more than once experienced the loss of much of a calf crop from range cattle because of a severe spring blizzard, I find not unreasonable the proposal that the absence of an entire age group from the 10,000-year-old Casper bison kill site could have resulted from a spring blizzard that had the same effect on a herd of bison (see Reher 1974: 115).

Severe winters must have tested the survival skills of prehistoric inhabitants. Some winters have moderate temperatures and light snowfall, while others are marked by deep snow and long periods of intense cold. These variations are totally unpredictable: several mild winters may follow several bad ones, or a single bad winter may occur in a string of several mild ones—or vice versa. Late-nineteenth- and early-twentieth-century livestock raisers suffered through a number of bad winters. Perhaps the most notable, and one I remember as a common topic of conversation a half century after its occurrence, was the winter of 1886 and 1887, which eliminated much of the livestock on the open range in Montana and Wyoming (see Larson 1978: 190–94) and, according to one rancher (now long deceased), killed the last nine bison known to have been in the Big Horn Basin. These winters were cynically referred to by livestock operators as "equalizer years," because having lost their economic base, everyone began the following year as equals. Livestock operators quickly learned to prepare for the worst and hope for the best. There is no reason to assume that these vicissitudes of weather were limited only to historic times, and prehistoric year-to-year human subsistence strategies had to allow for similar rapid and unpredictable changes.

One way to acquire a lasting impression of the region is to view it on clear days from the window of an airplane on flights up and down the spine of the Rocky Mountains from about Albuquerque, New Mexico, to Calgary, Alberta. Flights at all seasons of the year emphasize the contrasts, and shadows—particularly just after sunrise and before sunset—enhance the area's physiographic features. The Rocky Mountains might aptly be described as a ridgepole with major rivers flowing in all directions. The Rio Grande and the Colorado flow to the south, the Snake flows west into the Columbia, and the Missouri first flows northerly and gradually turns east. The Yellowstone flows east to the Missouri, as do both the North and South Platte. The Arkansas begins in the mountains in southern Colorado and flows east, eventually ending at the Mississippi. These river systems appear to have been avenues that many prehistoric groups followed in moving into the country, and they were critical to the overall prehistory of the area.

MAMMALS PAST AND PRESENT

The area described provided ecological conditions favorable for large grazers and browsers; as a result, it became the main focus of prehistoric large land mammal hunting in North America. The complete list of

TABLE I
Common Late Pleistocene and Holocene Mammals

Antilocapra americana	pronghorn
*Arctodus simus**	short-faced bear
*Bison antiquus**	bison
Bison bison	bison, plains bison
*Bison latifrons**	bison, giant Ice Age bison
*Bison occidentalis**	bison, western bison
Camelops sp.*	camel
*Canis dirus**	dire wolf
Canis latrans	coyote
Canis lupus	gray wolf
Castor canadensis	beaver
Cervus elaphus	elk
Cynomys sp.	prairie dog
Dicrostonyx torquatus†	collared lemming
*Equus conversidens**	Mexican horse
Erethizon dorsatum	porcupine
Euarctos americanus	black bear
*Felis atrox**	American lion
Felis canadensis	lynx
Felis concolor	mountain lion
Felis rufus	bobcat
Gulo luscus	wolverine
Lepus sp.	jackrabbit
*Mammuthus columbi**	Columbian mammoth
Marmota flaviventris	yellow-bellied marmot
Mephitis mephitis	skunk
*Miracinonyx trumani**	American cheetah
Neotoma cinerea	bushy-tailed wood rat
Ochotona princeps	pika
Odocoileus hemionus	mule deer
Odocoileus virginianus	white-tailed deer
Ondatra zibethicus	muskrat
Peromyscus moniculatus	deer mouse
Spermophilus richardsonii	Richardson's ground squirrel
Sylvilagus nuttallii	cottontail rabbit
Taxidea taxus	badger
Thomomys talpoides	northern pocket gopher
Ursus arctos	grizzly bear

*Extinct species.
†Extinct in the study area.

animals recovered in archaeological contexts is large, but the ones relevant to the following chapters consist of a relatively few species, some
extinct and others still present (table 1). The late Pleistocene mammoth
and bison are extinct, but the latter were apparently the progenitors of
the modern bison. Paleontologists seem to generally agree that there was
a late Pleistocene mountain sheep that gave rise to the modern species.
Pronghorn, deer, and elk are apparently unchanged from the earliest specimens known in archaeological contexts. The Pleistocene camel and horse
are both extinct, and the extent of their contribution to the Paleoindian
economy remains unclear. Gray wolves were eliminated but reintroduced
recently into Yellowstone National Park, where their numbers seem to
be rapidly increasing. It will be interesting to observe their future effects
on the wildlife there—particularly on the elk that, according to wildlife
managers, need a predator to control their numbers and prevent population growth that threatens their environment.

THE CULTURAL CHRONOLOGY

We need a chronological frame of reference within which to place the
human hunters and their animal procurement activities. The one used
here, which I developed and which applies over much of the area discussed (figure 1), is based on radiocarbon dates, stratigraphy, and projectile point typology. Approximately 5,000 years of this chronology fall
into divisions broadly labeled Plains Archaic. I concede that the word *Archaic* was poorly chosen, in that it denotes a lifeway not characteristic of
much of the area under consideration—as has been emphatically pointed
out (see, e.g., Forbis 1968, 1985). The term should be reserved for populations that practiced specific prehistoric subsistence strategies commonly
associated with that designation. However, I hope all readers will use the
chronology and overlook one term's awkwardness.

The following chapters contain my thoughts on prehistoric human
mammal hunters and hunting strategies as they pertain to our understanding of prehistoric hunters as seen in the archaeological record. I focus
on those sites pertinent to the topics of prehistoric animal procurement
and tool and weaponry use; they are not intended to be representative
of the overall prehistory of the area.

After long and serious deliberation, I define *hunting,* in brief, as the
exploitation of animals for specific purposes by human predators. Using
this definition as a point of departure opens the door to the exploration
of a voluminous body of information on the many different aspects of

YEARS BEFORE PRESENT	PROJECTILE POINT TYPES AND CULTURAL COMPLEXES	TIME PERIODS
12,000	Pre-Clovis ?	
11,500		
	Clovis	
11,000	Goshen	
	Folsom	
10,500	Midland	
	Agate Basin ?	PALEOINDIAN
10,000	Hell Gap ↑	
	Alder Complex	
9,500	Alberta |	
	Several unnamed	
9,000	Cody foothill-mountain	
	Paleoindian complexes	
8,500	Angostura ? Lovell Constricted	
	Frederick	
8,000	James Allen Pryor	
	Lusk Stemmed Complex	
7,500	Blackwater side-notched ↓	
7,000	↑	EARLY PLAINS ARCHAIC
	Pahaska side-notched	
6,500	Several unnamed "early side-	
	notched" and some corner-	
6,000	notched projectile point	
5,500	varieties	
5,000	Bitterroot	
	Oxbow ↑↓	
4,500	Mallory	
	McKean lanceolate	MIDDLE PLAINS ARCHAIC
4,000		
	McKean Complex	
3,500	Duncan and its variants	
	Hannah ↑	
3,000	Pelican Lake | ↓ ↑	
	Late Plains Archaic Yonkee	LATE PLAINS ARCHAIC
2,500	corner-notched point ↓	
2,000	varieties	
	Avonlea ↑	
1,500	Besant Late Prehistoric	
	↓ side-notched and	
1,000	corner-notched	LATE PREHISTORIC
	Prairie side-notched variants	
500	(some corner-notched variants)	
	Plains side-notched	
	(some corner-notched and	
	unnotched and many base-notched variants)	

FIGURE 1. Chronological chart and cultural complexes. (From Frison 1991b: 24.)

hunting. However, the following chapters focus on a limited topic—that of prehistoric large mammal hunting in western North America. They are not designed or intended to constitute a "how-to-do-it" list of instructions for either hunters or archaeologists: to make such a claim would raise the hackles of many in both groups who would immediately reject it and most likely decide against a further look at the pages that follow. The message intended is that a better understanding of prehistoric hunting strategies can be enhanced by those familiar with animals and their behavior in response to modern hunting strategies.

In retrospect, though I have been involved in hunting for more than seven decades I must admit to certain deficiencies. My proficiency with the bow and arrow never rose above average, and though somewhat more skilled with firearms I envied those who could repeatedly group their shots in the bull's-eye. My only recourse was to focus more attention on other hunter-related requirements to maintain an acceptable level of expertise. This leaves me with only the claim of fortuitously being in the right position at the right time to investigate a number of archaeological sites relating to large mammal hunting and being able to perceive that my experiences with animals, both wild and domestic, might benefit archaeologists faced with recovery, analysis, and interpretation of faunal assemblages resulting from prehistoric human hunting. With this caveat, I hope the reader will be encouraged to continue.

Some readers may conclude that much of chapter 2 wanders too far from my declared subject matter. I counter this with my own claim that more than once, seemingly insignificant observations of animals have long remained latent and suddenly materialized at the opportune moment to salvage what appeared to be a hunting strategy doomed to failure. In other words, while stalking an animal, the hunter not only needs to focus attention on the animal being pursued but, in addition, must be equally aware of the messages other animals, birds, and even insects are sending at the same time. In a much different vein, chapter 2 also portrays the interplay of influences that resulted in a combined interest in hunting, archaeology, and paleontology, which in turn led to an academic career.

My early interest in paleontology was fueled by a wealth of dinosaur fossils underfoot, but they never aroused the same level of interest as mammoth remains once I became aware that while they were said to be millions of years old, the mammoths, along with other extinct species, were only thousands of years old. This interest culminated much later in the discovery of a Clovis mammoth kill with several other extinct animals also represented. To me this is worthy of separate treatment, as presented in chapter 3, which explores late Pleistocene animal extinctions and the possible human relation to them.

Bison dominate the evidence of prehistoric large mammal hunting, which I believe justifies the depth and breadth of detail allotted to bison in chapter 4. Taphonomic studies of faunal assemblages in archaeological sites came into wide use when archaeologists were perceptive enough to realize that bone beds in bison kill sites had the potential to produce cultural information. That realization is tied to the underlying theme of this book: that animal and human behavior along with the evidence from taphonomic analysis reveal information significant for prehistoric animal procurement studies. I believe this approach has been vindicated during the past three decades, in large part through studies of bison bone beds.

Chapters 5, 6, and 7 rely on a similar approach to deal with the pronghorn, mountain sheep, deer, elk, bear, smaller animals, and birds whose skeletal remains occur in archaeological sites. Emphasis is placed on the contrasting behaviors of these species and the different procurement strategies that emerged as a result. I stress the advantages of being able to observe and pursue these species in their natural habitat and apply the information thereby gained to the evidence from archaeological sites.

Dispatching the animals is the final goal of hunting, and the manufacture, use, and maintenance of weapons determine the final coup de grâce (chapter 8). A hunter may stalk an animal in its natural environment and rely on his ability to outwit it; a cooperating hunting group may devise various ways to maneuver animals into favorable positions; in the case of buffalo jumps, varying numbers of animals were tricked into killing themselves. Even in the last case, weapons were needed to prevent the escape of animals that were not killed outright by the impact of their fall. In addition to killing animals, enhancing the quality and preventing loss of the products of the kill must be included as part of hunting. Consequently, tools as well as weapons must be included in discussions of hunting. Some archaeological site evidence indicates that projectile points were used as tools, a use that had some bearing on the value a hunter placed on them.

Chapter 9 presents some of the thoughts and impressions that have emerged as I compiled the preceding chapters. My overall sense is that although a significant body of data on prehistoric human hunting has accumulated over the past few decades, there is yet a long road ahead for researchers.

CHAPTER 2

The Education of a Hunter

THE YEARS BETWEEN THE WARS

Times were difficult in ranching communities in much of the Rocky Mountain West during the drought and depression days of the 1930s, and for many families wild game animals were an important supplement to the winter food supply. Wild animal populations—especially elk, deer, and pronghorn—were recovering from overexploitation at the turn of the century, and regulated hunting seasons were established. For many, the fall hunt became an integral part of the yearly cycle; the result was a generation of experienced hunters, a group concerned more with feeding their families than with garnering trophies to hang on the wall. In addition, predators were a constant problem, and hunting and trapping them was necessary to protect the livestock being raised, especially on the open range.

Firearms were always readily accessible but kept out of reach and strictly forbidden to children until they had received careful instruction. At about the age of seven, I was allowed the use of an old single-shot .22 caliber rifle. This privilege was bestowed with a set of definite and unbending rules. First, the firearm was to be loaded only when I intended to aim and shoot at something. Second, under no circumstances was I to point the firearm at any person. And third, any careless behavior with the firearm would bring its immediate confiscation until there was ample evidence that my attitude had changed. This last was a powerful incentive to abide by the rules.

Most ranch boys my age as well as many grown-ups ran a trapline during the winter, when hides were in prime condition. To catch coyotes, a dead cow or horse was cut into large sections and distributed at spots selected in the hills within a kilometer or two of the home ranch; two or more traps were set at each location. This was done mostly for predator control, but partly for profit as well. In those days, a prime coyote hide fetched $3.00; a bobcat hide, $2.00; a muskrat hide, $0.50; a jackrabbit hide, $0.10; and a cottontail hide, $0.05. Beaver pelts were valuable but rare, because the animals had been trapped out of this area: they would later be reestablished. Skunk hides had about the same value as those of muskrats, but skunks were nearly impossible to skin and process without the skinner taking on enough of their characteristic odor to be excluded from the family table at mealtime and sent home from school. The hide buyer usually came around twice during the winter, and a few dollars from the sale of hides was a real bonanza and about the only opportunity that kids my age and in my circumstances had to acquire real money. The .22 rifle was a necessity to dispatch a coyote or bobcat caught in a trap; in addition, there was always a good chance of picking off a rabbit or weasel, whose hide would add a few cents toward the purchase of a box of fifty cartridges (which then cost $0.35).

Trapping fur-bearing animals involved more than just setting a trap and waiting for an animal to blunder into it; the best way to learn about trapping was to follow an experienced trapper for a season or so and become familiar with the habits of the different animals sought. For example, mature coyotes were especially wary of metal traps. They usually circled a carcass used as bait at least once before approaching it, so traps had to be carefully concealed. The traps also had to be placed to minimize their attractiveness to birds such as magpies, ravens, and eagles, who might otherwise spring them. In addition, they needed to be in the shade: a crust formed by nighttime freezing of snow that had melted on a warm winter afternoon might support the weight of the animal and prevent the trap from springing. Many trappers used scent to cover up signs of a human presence; one common technique was to allow dead fish to ferment in a glass jar in the warm sun for several days. The fermented fish had to be kept well apart and downwind from where anyone was living or working.

A very different method was often successful in trapping bobcats: it relied on their curiosity. Something like a piece of dried hide or a bird's wing was suspended from a tree branch or from a peg in a dirt bank high enough to force the animal to stand on its hind feet to investigate. The trap was set below the bait and, for a bobcat, did not need to be concealed. Rock ledges and overhangs were ideal locations, because bobcats frequented these areas; other animals and birds had little or no interest in this kind of bait and so were unlikely to spring the traps.

Muskrats were trapped at entrances to their holes in riverbanks and swamps. Jackrabbits hid during the day, but in the deep winter snow at night they came into haystacks in droves for food and were easily taken in unconcealed traps. Cottontail rabbits were visible during the day and usually killed with the .22 caliber rifle. Simple snares at the entrance to burrows also worked well to trap cottontails.

There were a number of unwritten rules about trapping animals. Moving into another trapper's territory or taking animals from another person's traps was very definitely frowned on and could lead to serious conflict. Traps were to be monitored regularly to prevent the animals from suffering unnecessarily. Coyotes left in a trap too long would sometimes chew off a frozen foot; no longer able to follow their normal instinct to pursue wild animals for food, the three-legged animals turned their attention to domestic animals that were easier to catch and thus often became extremely troublesome.

As already mentioned, predators were an ever-present problem, and constant vigilance was needed to protect domestic animals and birds. A weasel could clean out the henhouse in a few minutes, and bull snakes loved eggs as well as newly hatched chicks. Bobcats and skunks were also attracted to henhouses. A coyote could quickly kill a grown sheep, carry off a newborn lamb, and occasionally kill a newborn calf. A female coyote that was teaching her pups to hunt in late summer and early fall could be especially bothersome. Raptors were true opportunists, always on the alert for unprotected birds and small animals.

Predator threats were often announced by distinctive distress calls from the intended victims; humans within earshot immediately responded by grabbing a rifle and rushing to the source of the disturbance, which could occur at almost any time of the day or night. I was a few years too young to witness the gray wolves that roamed the country in packs. Highly efficient predators, they were the major threat to livestock in earlier decades, but they had been eliminated in this area shortly after World War I. Their demise was hastened by the hefty bounties offered by livestock operators. One strategy to control both wolves and coyotes was to wait for a spring snowstorm, when a female could be tracked back to her den and the pups eliminated. My grandfather, an accomplished hunter and trapper, claimed a record of 40 gray wolves and more than 200 coyotes during one year shortly after the turn of the century.

Dogs were often a major problem for livestock owners. Some were unclaimed and managed to survive in the wild, while others were family dogs that formed packs and roamed the countryside at night. Any dog that appeared lethargic and slept most of the day was viewed with distrust; a dog that appeared to be a typical family pet when at home could take on an entirely different character when it joined other dogs. Food was usually not their main interest; they would chase and kill but seldom eat a dead animal. Domestic sheep were especially vulnerable, as were deer and pronghorn weakened by winter conditions. A domestic cow could usually but not always protect a newborn calf. Consequently, dogs running loose were commonly shot on sight.

Aside from an occasional visit, wandering grizzlies were rarely seen in the mountains where we lived; but sizable numbers of them were still present some distance away in hard-to-access areas in and around Yellowstone National Park. They were, and still are, the only species in the lower forty-eight states regarded as dangerous to hunt. They can move unexpectedly fast and can easily kill an animal as large as a domestic cow. Human-grizzly relationships have changed significantly since the latter were placed on the endangered species list. They are no longer hunted legally; to avoid paying heavy fines, anyone killing one has to demonstrate to the satisfaction of the authorities that he or she acted in self-defense or to protect livestock.

Black bear were usually not serious predators, preferring to scavenge dead animals for food; but if one became troublesome it was either shot or trapped using large double-spring traps, oversize versions of smaller coyote and wolf traps. Actually, the largest double-spring metal trap used for bear, known as the No. 6, was first designed for use on elephants. A bear, unlike a coyote, is not wary of metal traps. A large animal in an advanced stage of decomposition, whose smell could be detected over long distances, was the best bait for bear. An old horse about to die was a regular candidate for baiting a bear trap: in fact, any horse of substandard quality was commonly mocked as "bear bait." The late Roy Coleman of Cody, Wyoming, was an accomplished big game outfitter and grizzly hunter. He took me to his favorite spot, deep in the wilderness area outside the east boundary of Yellowstone National Park, where I was able to acquire grizzly skeletal parts for comparative studies. Grizzly hunters often took only the skull and hide, with claws attached, as trophies and abandoned the remainder of the carcass.

Mountain lions were present but scarce. They were highly efficient predators but preferred to stay in isolated areas of rough terrain, where they subsisted mainly by killing mule deer and small mammals. Because of their small numbers in those days, they rarely became a serious threat to livestock; thus little effort was made to trap them. Occasionally, however, a lion hunter with trained dogs would come into the area looking for a trophy. Within the past few decades, mountain lions have managed to increase dramatically in some areas and have become predators especially troublesome to deer, mountain sheep, and domestic sheep.

Though not predators, porcupines were a cause of some concern. Dogs trained for use around livestock are valuable property, and with few exceptions they learn the hard way that killing a porcupine will earn them a head—and sometimes other body parts as well—bristling with quills. One of the more unpleasant tasks that fell to those on a ranch was restraining a dog with a canvas or an old quilt and removing quills. Other animals also were potential victims: young colts especially tend to be overly inquisitive and get their noses too close to a porcupine, with the same inevitable results (colts' inquisitive nature sometimes results in rattlesnake-bitten noses as well). On two occasions I found carcasses of animals, one a colt and the other a yearling heifer, that had died because embedded porcupine quills eventually became infected and prevented them from eating. As a result, porcupines were usually killed on sight.

The outside observer unfamiliar with animals and hunting usually fails to comprehend the many decisions made by the hunter, who must use all of his or her senses to read the signs that reveal animal activities. There are almost endless examples. Tracks can reveal the species, number, sex, and age of the animals who left them, as well as the time and their mode of behavior when they passed through. Tracks indicate whether they were grazing and unaware of danger, frightened and pursued by predators, moving to or from water, or heading for protective cover. Fresh snow is always welcome, since it obliterates older signs and leaves an up-to-date record of recent animal movements. Both male elk and deer leave telltale signs by rubbing the bark off small trees to polish their antlers; male pronghorn paw the dirt, offering sure evidence of their presence nearby. Dust clouds can reveal the number, direction, and speed of animals moving in dry weather. Behavior outside animals' normal patterns of activity should never be ignored, as it may indicate other hunters or four-legged predators in the vicinity. Feces are also highly informative and can help indicate which plant species the animals are consuming. Armed with this knowledge, the hunter, familiar with the feeding patterns of the different species, can monitor areas where individuals are likely to be found during certain times of the day.

Sounds are important; the tone of the calls of many animals and birds sends messages about the activities of other animals. Mountain jays, magpies, crows, and pine squirrels emit special sounds that announce animals on the move; the bugle of a male elk is unmistakable and can be heard for distances of a kilometer or more and, under certain conditions, is answered by other males in the vicinity; cow elk emit a much different sound that can alert a hunter to their presence, especially in thick brush or timber. Mule deer bounding (stotting) at high speed produce a sound entirely different from that of a running white-tailed deer or pronghorn. And smell can also play a crucial role. Elk have a distinctive odor that can be detected for short distances under certain atmospheric conditions. Because they like to bed down in heavy brush and patches of timber during the day, this smell is a sure indicator of their location.

On the other hand, the animals' senses of sight, hearing, and smell are usually more acute than those of humans, and that keenness generally works to their advantage. Many hunts have been spoiled by careless talking, coughing, snapping off of dry branches, and tramping through dry vegetation. Mountain sheep and pronghorn rely strongly on eyesight to avoid predators. Bison and elephants (and possibly also mammoths) rely less on eyesight but have an exceptional ability to detect human odor.

This overview of sensory indicators used by both humans and animals is by no means exhaustive. Down through the years, the person who hunts and traps becomes so proficient at recognizing sights, sounds, and smells that his or her responses are automatic. The ability to best exploit clues available to the senses is a major determinant of success or failure. Most human hunting behavior that relies on such observations is subtle, almost invisible, and often overlooked by nonhunting observers.

WILD ANIMALS AND BIRDS HUNTED FOR FOOD

Cottontail rabbits were plentiful and occasionally hunted for food. Their popularity declined during the 1930s as fear spread of the rabbit disease tularemia, which can be transmitted to humans who have open cuts or sores and handle the flesh of infected animals. Jackrabbits rated even lower than cottontails as food; prairie dogs and marmots were rarely, if ever, eaten. About the only use for marmots was to catch one or two mature ones in the fall before they hibernated and render them for their fat, which was an excellent conditioner for saddles, chaps, and other leather goods. Porcupines, once located, are easy to kill with a weapon as simple as a club, but tradition holds that they were eaten only as a last resort, when the alternative was starvation. To test this folklore, I cooked and ate part of one; the results seemed to confirm earlier claims that porcupines rated extremely low as human food. On the other hand, Gilbert Wilson quotes his Hidatsa informant, who claimed that broiled porcupine was "very good, quite fat and very tasty" (G. Wilson 1924: 303). Other than an occasional sage hen, duck, or willow grouse, the animals available to be hunted to supplement the food stored for winter were deer, pronghorn, and elk. I always felt a bit cheated because of an absence of Canada geese where I lived. Their flyways lay both to the east and west; an occasional flock would fly over but only rarely stop during their spring and fall migrations.

Mountain sheep were regarded as excellent food, but they were so scarce that hunting them was severely restricted. According to accounts of fur trappers and early settlers, mountain sheep were plentiful around the turn of the century but their numbers dwindled from disease shortly afterward. It was believed to have been transmitted from domestic sheep, but the link has never satisfactorily been proven (see Honess and Frost 1942). Moose are recent migrants to the area, and there is no known record of them in local archaeological sites. The Wyoming moose, *Alces alces shirasi*, is a smaller version of the American moose, *Alces alces americana*. Moose are solitary animals and not difficult to hunt. The flesh of a young animal is good; that of a mature bull is less desirable.

In those days, most people tended to shy away from eating the flesh of carnivores and scavengers; I lost any desire to eat bear meat one warm spring day after watching a black bear eating a very ripe domestic cow carcass that was crawling with maggots. In addition, an old female that produced two cubs each year lived near our summer cow camp. She never threatened anyone or bothered the livestock, although she was something of a nuisance in raiding the garbage pit. Watching her and the cubs was a major source of summer entertainment, and hunting them would have been akin to killing a family pet. I shed a few tears myself when a hunter killed one of the cubs during the fall hunting season.

In retrospect, I could see that before I ever bought a hunting license, I had acquired a considerable body of knowledge about trapping and hunting wild animals. Deer were plentiful and with a lot of coaching from my grandfather, a dedicated and accomplished hunter, I killed my first deer at the age of nine and a pronghorn three years later. Hunting was more than simply learning how to pursue and kill animals. It was accompanied by a body of ethics relating to the treatment of wild animals and fellow hunters. The rules dictated that a hunter shoot at an animal only when the probability of a lethal result was very high. Long-distance shots, known as "desperation shooting," and shooting into a herd rather than selecting an individual animal were both definitely frowned on. Every possible effort had to be made to pursue and retrieve a wounded animal, and all edible parts of an animal were to be salvaged. If another hunter was ahead of you, you withdrew quietly and moved to another area to avoid animals he or she was stalking.

After further reflection on the past, I realized that my family, and many others as well, adopted an early form of environmental conservation. They stressed continually that the presence of wild animals improved the quality of life in general and that enough feed should always be left in the hills to ensure their winter survival. It was this kind of philosophy that aided in saving some species from extinction and hastened their recovery after earlier overexploitation.

Returning from a hunting trip with his or her first elk was an important milestone in the life of a young hunter. After tagging along on hunting trips beginning at about the age of six, at fourteen I was of legal age to purchase my first big game hunting license, miss two days of school, and participate in an elk hunt. Elk had been eliminated from our immediate area just after the turn of the century but were reintroduced shortly thereafter; by the late 1920s, the herds had increased sufficiently to allow hunting. In my eyes, a successful elk hunt would be the ultimate achievement. Visions of elk herds interfered with my sleep for several nights in anticipation of the big event. To hasten the passage of time, firearms were inspected, cleaned, and oiled repeatedly and hunting knives were honed to razor sharpness.

It was a day's ride by horseback into the elk-hunting area, and the hunting camp was an old abandoned homestead cabin in the mountains several kilometers from the home ranch. The roof leaked and the chinking between the logs was gone, but it boasted an old iron stove for heat and cooking and was a welcome alternative to a tent camp. A nearby makeshift barn had collapsed, so the horses were tied in a thicket of lodgepole pine that afforded some protection; they were given an extra ration of grain to make up for a lack of hay or pasture. It was bitter cold and still dark the next morning when we saddled the horses and left camp to hunt. I was concentrating more on keeping warm than paying attention and looking for elk when just after daylight one of the hunting party yelled "There they go!" By the time I got off my horse, pulled the rifle out of the scabbard, and levered a cartridge into the chamber, the hunt was over. Two elk were dead and the rest of the herd was well out of rifle range and disappearing over the next ridge. Two more days of hunting produced nothing, and it was time to return home. That year, I would have to be satisfied with deer and pronghorn hunting. It was disappointing not to get an elk and more than obvious that I had a lot yet to learn. However, I was determined that things would be different next time.

The following year my uncle, an experienced elk hunter, and I waited until the late part of the hunting season, hoping for heavy snow to bunch up the elk and force them out of the high country and down into areas where they would be more accessible. As we arrived in the hunting area, subtle changes in our horses' behavior led us to expect a change in the weather. In addition, two bunches of mule deer were stirring and beginning to move downslope toward lower elevations, nearly always a reliable sign of an approaching storm system at that time of year.

After we set up camp, a quick reconnaissance just before dark revealed a small herd of elk about 2 kilometers away that gave all indications of bedding down for the night. Clouds were moving in: it would be a dark night with no moon. Under such conditions, we expected the elk to stay bedded down and not be up grazing. A careful check indicated that no other hunters were present who might have designs on this same herd. We believed we could get close enough to them at daybreak for a good chance of success.

The next morning began a cold and gray late fall day with an occasional snowflake in the air and the promise of heavier snow later. Leaving our horses tied in a patch of timber, we proceeded on foot to where we had spotted the animals. However, the wind shifted unexpectedly; by the time it was light enough to see, the elk had detected our scent and were disappearing over the next ridge, headed for a large patch of thick pine timber. Pursuit at that time would have been futile. It seemed better to return to camp and hope they would emerge from the timber late in the afternoon. Our predictions were good, but they emerged too late in the day and too far away for us to get close enough before dark. It was back to camp again, cold, hungry, and disappointed, to map out plans for the next day.

The animals were undoubtedly aware of a human presence but apparently had not been closely pursued or shot at recently; they gave no indication of moving to another area, as elk commonly do when heavily pressured by hunters. Under these conditions, another try the next morning was well worth it. At first good light, none of the animals were in sight. Several inches of fresh snow had accumulated during the night, and we soon crossed their tracks. They were moving slowly and grazing in an area of scattered trees and tall sagebrush, apparently unconcerned and unaware of any danger. The wind was in our favor, and after a short stalk, we found feces not yet frozen in the snow—a sure indication they were not far ahead of us. By early afternoon we spotted them before they spotted us and, with that advantage, we were able to secure a position within good shooting distance. Winter rations and not trophies were our major concern, and we were able to get close enough to select what we believed would be the best meat. One elk apiece was allowed, and my uncle and I each made clean shots on a young bull and a barren cow, both in excellent condition, guaranteeing that we would have top-quality meat for a good share of the coming winter.

It was too late in the day to pack the animals to camp, but the weather was ideal for cooling the animals. Both were field dressed, and green branches were placed over them to discourage scavenging by the ever-present magpies, ravens, and eagles. We could detect no evidence in the immediate area of bear that might welcome a free meal. Coyotes were around, but they seldom bother a carcass the first night; as an added precaution, we moved the innards a short distance away to serve as a decoy. Hearts and livers were taken to camp. This time we had smiles on our faces and the prospect of fresh liver for breakfast. After breakfast, we had a long, hard day's work, but we had our best hunting horses, which were used to carrying dead animals. Each animal was split down the middle with a sharp axe; since each one weighed about 400 pounds, half an elk was loaded on each horse. Returning to the home ranch the next day, we didn't have to sneak around the corner of the barn and slip inside to unsaddle the horses and later face the rest of the family with lame excuses for why we claimed to be good hunters but were unable to outwit the elk.

In the late 1930s, refrigeration on the home ranch was still several years down the road, but the weather was now cold enough to hang the meat in the shade with the hide intact without risking damage from spoilage or flies. It would be skinned out and used as needed. Elk meat was regarded by most as the best wild game, with a flavor at least equal to that of beef. Deer and pronghorn, which have their individual distinctive flavors, were more plentiful and easier to acquire than elk. This elk hunt was the culmination of a very successful hunting year.

After that, the coming long winter was filled with school and chores. Nothing could equal or surpass the personal satisfaction gained from that fall hunt. I was finally accepted by elders and peers alike as one who was becoming a full-fledged hunter and provider. Times may have been tough but I never felt deprived; in fact, I thought I was about the luckiest kid on earth. In reality, it was only the beginning of a hunting lifeway, because every subsequent hunting episode added to my body of knowledge of how to go about animal procurement. In the following years I developed special feelings toward game animals, especially elk. I gradually came to understand something of the spiritual relationships that can develop between subsistence hunters and the animals they kill for food.

Hunting was quite different then from what it became even a very few years later. Our firearms were the traditional lever-action Winchester Model 94, .30–30 caliber rifles with open iron sights that most hunters openly sneer at today—not the telescopic sights and flat-trajectory, high-velocity calibers in use a short time later. There were no four-wheel-drive vehicles, and access to remote hunting areas was mostly on foot or horseback. Binoculars were a luxury then, and one that we didn't have. A box of cartridges was too expensive to be wasted on long shots, on animals moving through timber or brush, or in other situations in which the chances of delivering lethal wounds were remote. Instead, the hunter was expected to use his or her experience and knowledge of the territory and animal behavior to get into a position to raise the probability of success.

At the same time, hunters are a competitive and arrogant bunch. It became common practice to embark on hunting trips with no provisions other than salt and coffee, trusting that an animal would be killed for camp use. A common expression at the time, which held an element of truth, was that nothing sharpened hunting expertise as quickly as hunger. This hunting era ended abruptly with the events at Pearl Harbor on December 7, 1941. Nearly everyone around my age either enlisted or was drafted into the armed forces soon afterward.

THE POSTWAR PERIOD

After World War II was over, hunting conditions began to change. Hunting pressure was low during the war years, and wild animal populations increased. Nearly everyone was now more affluent and anxious to travel, and many sought new forms of entertainment. When word spread about all the game animals in the Rocky Mountain states, hunters flocked there by the thousands. Returning servicemen had acquired military firearms of every description, along with the latest in optical equipment, and insulated clothing made life in the field more pleasant. Some began to use war surplus four-wheel-drive vehicles rather than horses to reach remote camps. There were few restrictions on access to either private or public lands, allowing this new crop of hunters to saturate hunting areas.

Success rates were high for a few years, and hunting began to become frenzied as ethical standards concerning the treatment of landowners, fellow hunters, and the game animals rapidly fell. For example, in what became known as "the firing line" near Gardiner, Montana, masses of hunters congregated to wait for the large elk herds that migrated north out of Yellowstone National Park ahead of the heavy winter snows. As an elk herd left the protection of the park, the rifle fire sounded like a major military engagement. As that herd moved out of range and the firing stopped, the hunters, many wearing running shoes and not hampered with rifles, rushed out to claim the dead animals before the next herd appeared and the shooting started up again. Blood trails of the wounded were largely ignored, and many animals were left to wander off and die. A law requiring each hunter to carry a firearm was finally passed. Similar slaughters of elk, but of less spectacular proportions, occurred in the open flats adjacent to the National Elk Refuge in Jackson Hole, Wyoming.

Populations of small towns in hunting areas mushroomed before and during the hunting seasons. Hunting camps were everywhere, made more livable with new conveniences that included propane tanks and electric generators to supply lights and refrigeration. Dead animals, especially those with large and impressive antlers or horns, were tied on vehicles in the most conspicuous spots possible for everyone to see and admire. Many prime animals spoiled through neglect or ignorance of proper care in the field. Even though heavy penalties were levied on hunters caught taking only the heads of animals and abandoning the rest of the carcasses, the practice was all too common. It was the beginning of the period when big game hunting developed a very widespread bad public image.

Everyone knew it was now cheaper to buy meat at the butcher shop than it was to underwrite the cost of an extended hunting trip, so the economic justification of wild game hunting became less and less tenable. To be sure, most hunters did try to take care of their animal products responsibly. Wild-game-processing plants became viable, short-term, fall business ventures: anyone proficient as a butcher could take on part-time work skinning and cutting up the piles of dead animals that accumulated. But most of this new crop of hunters wanted something to show the folks back home. Even though they valued the meat, better evidence of their hunting successes was trophy animal heads; as a result, guiding and outfitting enterprises with the facilities and expertise to escort hunters into areas with trophy animals began to flourish. Taxidermy shops took on new importance, because many of these new hunters wanted their trophies enhanced for more impressive display.

THE TROPHY HUNTERS

In a long-standing tradition, begun in the late nineteenth century and involving a small number of individuals from both the United States and abroad, prolonged hunting trips are undertaken for the sole purpose of acquiring outstanding big game heads to stock trophy rooms or to adorn the walls of homes, clubs, and public buildings. They call to mind the nineteenth-century trips to biblical lands by wealthy Europeans to collect archaeological treasures (see Daniel 1952: chaps. 1–2). Large retinues accompanied the early trophy hunters, who demanded and received special treatment to alleviate harsh conditions in the field. Perhaps the best known of these personalities in this country was Theodore Roosevelt, whose hunting exploits around the world led to the killing of very large numbers of big game animals selected for their outstanding sets of antlers and horns. He has been excused for many of these excesses because of his positive efforts in establishing national parks, national forests, and wildlife refuges.

In part to establish bragging rights over who bagged the biggest and best trophies, the Boone and Crockett Club, named for two noted American frontiersmen, was founded (by Roosevelt). It developed a system of standard measurements for comparing the heads of big game animals: measurements are taken of antlers and horns and the final tally in inches establishes the ranking of that particular animal in its species. Status and, in many cases, even monetary awards accrue to the owner of an unusually fine trophy. Competition is keen among those wishing to see their names high up on the Boone and Crockett register of trophy animal heads (see, e.g., Webb et al. 1952; Byers and Bettas 1999).

Experience soon teaches the aspiring hunter that successful hunting is closely tied to familiarity with a hunting territory and with how its resident animals exploit it. Although there are expansive areas of state and federal lands in the Rocky Mountain states open to all, hunters with limited time who had to travel far and spend much money soon learned that the quickest and easiest way to ensure a successful hunt was to employ a local guide who knew both the country and the location of trophy animals. I soon realized that guiding hunters could provide me with welcome additional income to supplement a modest livestock ranching operation, and I began accepting a limited number of out-of-state hunters each fall. I knew the territory where the animals could be found and became quite successful in sending hunters home with trophy animals. It was, however, an entirely different approach to hunting than the one I was familiar with. The trophy seeker needs only to aim and pull the trigger to verify his status as a "hunter," relying on the guide to lead him to the proper spot and point out the animal with the best potential to rate high on the Boone and Crockett list. Quality or quantity of meat is rarely a consideration.

The trend toward trophy hunting gradually changed wild animal populations. The demand for larger sets of horns and antlers reduced the number of breeding males, so that females soon far outnumbered them; that imbalance, in turn, led to the establishment of hunting quotas. There were still local hunters willing to harvest females and young animals for food, although these so-called meat hunters were clearly looked down on by the trophy hunters. As hunting pressure became more intense every year, animal populations went into a decline; limits were therefore set on license sales. However, there were still enough relatively inaccessible wilderness and national forest areas to support trophy hunting.

Further support for the trophy hunters was provided by laws requiring out-of-state hunters to hire a licensed guide in wilderness areas. Many guides and outfitters had unwritten agreements to share hunting territories and resented any intrusions. This situation still prevails to some extent, because the average hunter lacks the means to successfully exploit wilderness areas and some of the more inaccessible national forest areas. Motorized equipment of any kind is prohibited from wilderness areas, a ban that also excludes many hunters unless they have horses or are willing to go on foot. Guide and outfitter licensing requires a considerable investment in animals, equipment, and facilities to accommodate hunters in the backcountry. Before his or her license is issued, a guide or outfitter must post surety bonds to ensure observance of game laws. Both outfitter and hunter may forfeit their licenses if caught violating those laws.

Mountain sheep, moose, and elk are the mainstay of the professional outfitting business in wilderness areas, while in many regions mountain lions have demonstrated a remarkable numerical increase and rank high as trophies. The grizzly bear was highly desired as a trophy animal until it was put on the endangered species list, where it still remains. Per-

mits for these animals are limited, and those who are allowed to hunt either moose or mountain sheep must by law wait five years before making another application, regardless of the outcome of their hunt. Moreover, because permits are nontransferable a successful applicant cannot
sell his or her permit to the highest bidder. These are strong incentives
for the permit holder to hire a reputable guide to increase the likelihood
of success: the hunt may be the only chance of a lifetime. While large
deer and pronghorn heads are also highly valued trophies, these animals
are present outside the wilderness areas and on private land, where licensed guides are not required. In addition, applications for permits to
hunt them can be submitted every year.

There is a dark side to the acquisition of trophy animal heads: a black
market serving collectors who operate outside the legally regulated hunting process. National parks, state parks, and wildlife refuges are home
to many of the best trophy animals. The very high prices they can bring
encourage poachers to kill these animals for their heads alone and to leave
the remainder of the carcasses behind. Some poaching is highly organized,
even to the use of helicopters to access trophy animals in remote areas,
few of which have adequate law enforcement. And because biological
collections in museums and research institutions are also targeted by
thieves, animal heads now demand tight security. A trophy head that
ranks high on the Boone and Crockett list can command a dollar price
well into the five-digit range.

Although I have here emphasized the negative aspects of present-day
trophy hunting, the desire for trophies is to some extent part of the philosophy of nearly every hunter. It is difficult to discard an outstanding
set of antlers or horns. Nailed to the side of the barn, on a post near the
front door, or to the ridgepole over a house's main entrance, it sends a
message to all visitors, establishes a bond with other hunters, and is a
constant reminder of past successes. There is some evidence of this kind
of human behavior found in archaeological contexts—for example, large
mountain sheep heads placed in trees near trapping complexes.

HUNTING AND AVOCATIONAL ARCHAEOLOGY

Along with a passion for hunting, I early developed an interest in archaeology. It would be difficult to find an area better endowed with evidence of prehistoric hunting than the plains and mountains where I was
raised. Canyon walls and other rock faces are adorned with pecked,
painted, and incised figures of bison, elk, mountain sheep, pronghorn,

FIGURE 2. Prehistoric cave painting of a bear, a bison, and a human form. (From Frison 1978: 413.)

bears, and birds, as well as anthropomorphic and geometric figures, left by earlier inhabitants (figure 2). Many of the animal figures are portrayed with arrows penetrating their bodies, strongly suggesting that ritualized relationships were recognized between animals and their human hunters. Mountain sheep and elk heads of trophy size have been found where they had been placed on natural shelves in rock shelters and forks of trees; some show the deliberate opening of brain cavities, yet another suggestion of ritual activity. Remains of an immense mound of badly deteriorated, shed antlers of elk, now completely weathered away, lay in a place where large elk herds wintered in the past and still winter today. They were heaped there before any recorded history, almost certainly by Native Americans, as were other large elk antler piles in the region (see N. Nelson 1942; Denig 1930).

A well-marked trail connecting early historic Native American camps on the Yellowstone River in Montana with the North Platte River to the south in central Wyoming traversed several kilometers of the old home ranch on the western slopes of the Big Horn Mountains. From the early age when I was first able to handle a horse alone, I spent many days riding along this trail and picking up broken and discarded horse travois poles, tipi poles, horse gear, an occasional metal projectile or knife, and numerous other items. Native American burials in rock crevices and on platforms in trees were still in evidence; some of those interred were wrapped in bison robes, others in European-made blankets. They were usually accompanied with personal possessions; these sometimes included bows and arrows and occasionally a horse killed at the site. In one crevice burial was a badly deteriorated leather garment with more than 100 pairs of elk incisor teeth attached to it. In another were the metal parts of an old flintlock firearm with the gunflint still in place. Still another contained a metal knife sharpened on one side, with the cutting edge serrated in a manner reminiscent of earlier flaked stone knives.

A neighbor found a sinew-backed bow and a bundle of twenty-seven arrows with a burial in a crevice protected from the weather. All were surprisingly well preserved except that the end of the bow buried in sand was somewhat deteriorated and all the sinew backing was cracked and curled. Most of the arrow shafts were tipped with stone points; a few were of hammered iron. I was able to keep the bow long enough to make replicas, one of chokecherry and one of juniper, but without the sinew binding, whose significance it took me some time to realize. It was longer yet before I could learn how to prepare and apply it. Nevertheless, within a relatively short period, I was able to gain enough expertise with the weapon to kill rabbits, prairie dogs, and marmots, using both stone and metal points. The chokecherry wood bow gave the best results, though lacking a sinew binding.

Physical evidence of the past presence of bison was everywhere. Their bones were eroding out of arroyo banks, and partial and complete skeletons survived in rock shelters and other protected locations. I remember seeing a metal projectile sticking in the horn sheath on one bison skull and finding another skull with a round hole in the frontal bone. On picking it up, I discovered a rattling noise inside that indicated the source of the hole: a lead musket ball. The frontal parts of many skulls were smashed or chopped open to expose the brain cavity. In a higher altitude grassland setting, an old settler pointed out numerous circular, shallow depressions that were bison mud wallows frequented as late as the early 1880s. Several years later, this same informant showed me the scattered skeletal remains of several bison that he claimed were the last ones remaining in the Big Horn Basin of northern Wyoming, which had perished in the disastrous winter of 1886–87 along with nearly all of the region's domestic cattle.

A never-to-be-forgotten incident with living bison occurred in the summer of 1935. Bison were scarce, but the state of Wyoming maintained a small herd in a state park near Thermopolis, about 100 kilometers from our home ranch. Present-day bison occasionally revert to their old habits: they have strong urges to wander at times, and unless they are restrained behind high, sturdy fences, their wanderings are difficult to stop. Fences on the open range were few and far between until World War II, and a herd of ten bison left the state park; they eventually found their way to our mountain pasture and, as bison are prone to do, settled down in an area of good grass and water.

I was allowed to tag along, but given strongly worded instruction to stay out of the way, when three local cowhands were hired to escort the animals back to their home range. One animal, a young bull, refused to cooperate and created so much confusion among the other animals that he was left behind. One day I tried to outrun this young bull on horseback. Approaching a barbed wire fence, he easily cleared it without slowing down. The same maneuver again produced identical results. On the third attempt, however, and at the last instant, he refused to jump the fence, executed a 180-degree turn, and ran between my horse's front legs, depositing both of us on the ground. Neither of us was hurt but this was the first among my many lessons in dealing with bison. They may appear docile and even awkward until prodded into action, at which time they can outrun and outmaneuver the average saddle horse.

For some time, I had been under other influences relating to archaeology and paleontology. Paleontologists were attracted to the fossil-rich Big Horn Basin in northern Wyoming. Barnum Brown of the American Museum of Natural History, who identified the bison at the Folsom site in New Mexico (Figgins 1927), excavated in dinosaur beds close to our ranch in 1934. He was a gracious host and tolerant of a nine-year-old boy with two cigar boxes of fossils and artifacts; while looking at my collection he mentioned Folsom points, about which I knew nothing. He would have been surprised to know that he was only a short distance from the Hanson site, one of the larger Folsom sites on the Great Plains, which I excavated between 1973 and 1975 (Frison and Bradley 1980). I never saw Brown again after his 1934 excavations in northern Wyoming, but later on I learned that subsequent to identifying the Folsom site bison, he (1932) investigated Late Prehistoric–age buffalo jumps along the Yellowstone River in southern Montana.

Harold Cook, the geologist from the Agate Fossil Beds National Monument in western Nebraska, stopped by our ranch one day looking for new fossil beds. On learning of my interest in archaeology, Cook related his experiences in establishing the integrity of the Finley Paleoindian site in western Wyoming and arranging for the University of Pennsylvania to become involved in the excavations (Howard 1943; Moss et al. 1951). Although the unexcavated parts of the Finley site were subsequently looted, faunal materials salvaged more than two decades later added new information on the bison population involved (Haspel and Frison 1987; Todd, Rapson, and Hofman 1996).

In the 1930s and 1940s, Glenn Jepsen, a well-known paleontologist from Princeton University, conducted annual excavations for Paleocene-age fossils in the Big Horn Basin at another site not too far from us. He, along with researchers from the University of Pennsylvania, became involved in the late 1940s in the excavation of the Horner Paleoindian site near Cody, Wyoming (Jepsen 1953), and Waldo Wedel of the Smithsonian Institution joined the investigation during its second year. The Horner site received national news coverage, and its list of visitors included most of those involved in North American Paleoindian (then known as "Early Man") studies at that time. One of these was Kirk Bryan, who studied the Quaternary geology of the Colorado Front Range in the 1930s in attempts to date the Lindenmeier Folsom site, which was then being investigated by Frank H. H. Roberts of the Smithsonian Institution (Roberts 1935, 1936; Bryan and Ray 1940). Unfortunately, Bryan died of a heart attack on the first night of his visit to the Horner site. Jepsen himself died before he was able to complete his analysis of the Horner site. Shortly after his death, I was fortunate enough to analyze the site assemblages recovered in the early excavations, conduct new excavations, and publish the results (Frison and Todd 1987). The Horner site became known as the type site of the Cody Paleoindian Cultural Complex because it produced the Eden- and Scottsbluff-type projectile points and the Cody-type knife together in a single site.

In the fall of 1952 I made a discovery that opened the door to a better understanding of prehistoric hunting. While using binoculars to locate deer, I spotted the opening of a cave (figure 3). It was across a deep canyon and too far away for me to try to gain access at the time; I did not return to closely investigate the location until the next summer. It was a relatively large cave, and the front part was dry with several pack rat middens in recesses of the cave wall. In one of these were several parts of wooden shafts covered with red pigment. Several other wooden objects were protruding from the edge of a large rockfall. Unsure of what they represented, I took them to William Mulloy, a recently hired anthropologist at the University of Wyoming. One item proved to be the distal end of an atlatl with spur intact, and another was the distal end

FIGURE 3. Entrance to Spring Creek Cave, northern Wyoming. (From Frison 1991b: 107.)

of a dart mainshaft. Still another piece was the cone-shaped proximal end of a foreshaft that fit into the hole in the end of the mainshaft.

At the time, I knew very little about atlatls and darts, but I was able to obtain copies of C. B. Cosgrove's report (1947) on similar perishable materials from caves on the Upper Gila and Hueco areas of New Mexico and Texas as well as M. R. Harrington's report (1933) on perishable dart shafts from Gypsum Cave near Las Vegas, Nevada. Using their descriptions and the parts from the cave in the Big Horn Mountains, I was able to make replicas of the weapons; over a period of about three years, I managed to gain enough proficiency to hunt rabbits and prairie dogs. I preferred the bow and arrow, which I found provided greater accuracy, over the atlatl and dart. The cave site, known as Spring Creek, ultimately yielded a large assemblage of wooden, stone, and bone artifacts that was later described and published (see Frison 1965).

For a number of reasons, the family ranch was sold in 1962, and my family's combined livestock and big game outfitting and guiding business ended. It was a difficult decision, somewhat eased by the growing emphasis on trophy hunting, which destroyed too many of the rewards I had enjoyed when hunting earlier under a much different set of conditions.

After more than two decades of avocational interest in archaeology, I had convinced myself that my years of rubbing elbows with archaeologists, paleontologists, and geologists enabled me to claim some sort of professional status. At a meeting of the American Association for the Advancement of Science in Denver in 1961, several of the professionals backed me into a corner and made it clear that to have any kind of future as an archaeologist, I needed academic degrees. I took their advice seriously, and the result was a sojourn in undergraduate and graduate school as a nontraditional student starting at age thirty-seven. It was an eye-opening experience, to say the least, but I have never regretted my decision.

HUNTING AND ANTHROPOLOGY

Anthropological studies offered me a new and different perspective on human hunters and hunting. As a graduate student, I felt that the approach taken toward hunting in the anthropological literature seriously lacked credibility. I found most ethnographic accounts of Native American hunters, along with accounts of early explorers and fur trappers, informative but rarely presented in ways that took into account the hunted animals' behavior. For example, the 1957 film *The Hunters* (directed by John Marshall), portraying the obviously staged pursuit and killing of a giraffe by Bushmen, has been shown to anthropology classes for decades. Like the ethnographic accounts, it is informative but violates too many rules of intelligent pursuit and killing of animals; paramount among these are to avoid taking unwise and unnecessary chances with a long shot at a running animal and not to pursue a wounded animal so closely that it cannot lie down. Often, even a mortally wounded animal is able to summon up enough adrenaline to travel a long distance. Most unbelievable, in my view, was the film's end, when the wounded but still upright and staggering giraffe is harassed by its weapon-wielding pursuers until it finally collapses in a heap; several of those watching (myself included) proposed that it was ultimately dispatched by a shot from a rifle, a suggestion later denied. Little in the entire episode conveys an authentic sense of an effort by experienced hunters.

I was also astonished, in a more positive vein, to learn of the ritual sanctions placed on hunting by members within their own social groups. Although I became convinced that most ethnographers and early travelers in the West lacked enough experience with animals to understand the finer details of hunting, their accounts do reveal special relationships between human hunters and the animals they kill for food. These data tell us that ritual and supernatural beliefs forced many restrictions on historic Plains Indian hunting activities, often preventing the hunters from taking full advantage of their hunting skills. Hunting large mammals successfully was and still is an activity with a high incidence of failure. Consequently the shaman was an important figure, relied on to invoke supernatural help. This reliance is well portrayed by Julian Steward (1938) for communal Shoshonean pronghorn hunting; by John Ewers (1949) and David Mandelbaum (1940) for the bison-hunting Blackfoot and Plains Cree, respectively; and by Robert Spencer (1959) for the North Alaskan Eskimo who hunt large Arctic mammals.

Supernatural restraints imposed on band and tribal hunting groups revolve mainly around beliefs in animal spirits and their relationships with humans. The underlying principle is that animals allow themselves to be taken by humans, but only as long as the animal spirits receive proper treatment. If hunters fail to observe this rule, the animals will not make themselves available and the entire human group will suffer (see, e.g., H. Driver 1961: 71–83). Once there is the perception within the hunting society that lack of success is caused by the hunter's failure to placate the animal spirits, religious specialists or shamans can inject a strong element of control over the procurement process that can hamper the hunter's true ability.

Somewhat different restraints are related to general principles of primitive economics (see Sahlins 1972). There was nothing to be gained by killing more animals than needed: distribution of meat products was largely determined beforehand, surpluses could not be sold or bartered for profit, and storage for future use was limited. True hunters rarely indulge in the wanton killing of animals. I have guided hunters who became so excited at the moment of the kill that they lost control and killed more than the one animal they were allowed, but this was the exception rather than the rule. Whether the victim is a domestic animal or one in the wild, a person nearly always feels regret at taking an animal's life. I make this declaration fully aware of historical accounts of bison hide hunters dropping animals as long as any remained standing; hunters aboard trains shooting bison out of open windows; and hunters, realizing the bison would soon be a thing of the past, frantically rushing to bison country to be able to claim they killed one of the few remaining specimens (see Haines 1970: 178–207; McHugh 1972: 247–70). I find it difficult to find a category into which to place this kind of killing of animals.

Another textbook approach to interpreting prehistoric hunting relies on artists, most of whom know little about the subject matter and whose renditions of prehistoric hunting episodes are therefore often fanciful and unrealistic. They portray large mammals chased into bogs, leaping over cliffs, driven by wildfires, and dealt lethal blows by thrown rocks. At the same time, dead and crippled hunters are being dragged from the scene. Most also leave the erroneous impression that wild animals make little effort to evade human predators and that delivering lethal wounds with primitive weaponry required only that the hunter approach within a short distance of an animal, and then thrust a spear or launch a dart with an atlatl; the target animal seems to regard the entire operation with detachment, apparently embarrassed to be involved in such a ridiculous episode. In reality, and as all experienced hunters know, wild animals quickly become used to a human presence but a single act of aggression on the part of a hunter can incite them to revert to their wild and defensive behavior (see Berger, Swenson, and Persson 2001). The early impressions made by such fabrications follow students throughout their careers and can lead them to erroneously interpret the archaeological record.

FAUNAL STUDIES, ANIMAL BEHAVIOR STUDIES, AND ARCHAEOLOGY

Bison were the predominant food animals killed in prehistoric times on the Great Plains and in the Rocky Mountains. After my many different kinds of exposure over several decades to bison, both living and dead, in figures on rock faces and in written accounts, it was no accident that my first ventures into field archaeology as a graduate student would be investigating bison jump sites (see Frison 1967a). I then excavated an entirely different method of bison procurement, an arroyo bison trap (Frison 1968b), and later a campsite with a large assemblage of pronghorn bones, believed to have been near a pronghorn kill (Frison 1971b). By this time it was becoming more and more obvious to me that the existing methods of excavation, analysis, and interpretation of faunal remains were producing quantities of redundant data and relatively little information about prehistoric hunting and hunters.

Other archaeologists were beginning to think along the same lines; several landmark studies appeared, including investigations of the Old Women's Buffalo Jump in Alberta (Forbis 1962a), the Boarding School Bison Drive in Montana (Kehoe 1967), and the Olsen-Chubbuck Paleoindian bison kill site in eastern Colorado (Wheat 1967, 1972). In these reports one could see the idea emerging that more innovative recovery and analyses of faunal remains in archaeological sites had the potential to reveal past cultural activities as well as biological and taxonomic information on the animals. Taking a new look at faunal materials would revitalize archaeological studies of prehistoric hunters.

When I was hired by the University of Wyoming, I gained an unexpected opportunity to pursue research in archaeological faunal studies at an institution close to both prehistoric animal kill sites and contemporary large wild animal populations. Nearby were located several bison herds living under open range conditions similar to those of a century earlier. These were more useful for research than bison raised in small, fenced pastures. In addition, their owners offered full cooperation in research projects. Elk, deer, and pronghorn populations, along with large predators, were literally at my back door and available for animal behavior studies. Thus I also had opportunities to accumulate the comparative skeletal collections, including animals of all ages and both sexes, that are indispensable for reliably identifying faunal remains from archaeological sites.

Archaeologists should acknowledge the paleontologists who brought the science of taphonomy into their field (see Efremov 1940; Voorhies 1969). Taphonomic analysis deals with everything that affected paleontological materials from the time they were part of a living population until they were exhumed from the ground by the investigator, and archaeologists rapidly perceived its utility for their examination of faunal assemblages (see Lyman 1982). Taphonomic analysis provides an ever-expanding source of new information, as the new methodologies developed through innovative research appear to be endless. Depending on how the material was recovered, even faunal assemblages collected before the development of the present methodology can often be analyzed in accordance with many principles of taphonomy. It thus has become possible to calculate both the amount of economic resources derived from hunting and the seasonality of specific events. The results offer a better chance of accurately interpreting the lifeways of prehistoric hunting groups.

Such progress has been made possible by a strong interdisciplinary approach, although the relevant data usually lie on the fringes rather than in the mainstream of other disciplines. This reliance on the fringes has often been a barrier to interdisciplinary cooperation and communication, but the future looks encouraging as new subdisciplines such as geoarchaeology and bioarchaeology are now emerging and gaining recognition.

THE WANING YEARS OF A HUNTER

The best years of hunting, like most activities that require strenuous effort, are limited. But the education of a hunter never ceases, and increasing age brings new perspectives that allow some successful modification of old hunting habits. As knee joints and other body parts deteriorate and reaction time lengthens, wisdom acquired through long experience can even the odds for a few years; but younger generations of hunters are always on the sidelines ready and eager to challenge their elders, and youth and exuberance inevitably win out in the end. It is difficult to swallow one's pride and be forced to accept meat from an animal killed by another hunter. I imagine a caricature—a man too old and feeble to hunt, desperately trying to save face as he points to the ground and attempts to explain to a group of impatient and bored youngsters how to interpret animal tracks and feces.

The aging hunter finds it increasingly difficult to climb steep slopes: rather than going directly up a steep incline, I choose a winding route that provides an easier grade. Brush patches and uneven terrain are avoided if possible, and quick bursts of running are no longer an option. Care is taken to locate obstacles such as rocks, tree limbs, and depressions that might cause bad knees to give way and lead to serious injury. To mount a horse, I take advantage of uphill slopes, tree stumps, rocks, or anything available in order to bend a knee enough to get a foot into the stirrup. Extreme caution is necessary on snow and ice; traveling in deep snow rapidly exhausts my store of energy. Because my reflexes have slowed, an animal erupting from a thick brush patch more often than not makes its escape before I can respond. As with my first elk hunting experience but for different reasons, the animals are usually long gone by the time I can dismount and get organized. A better strategy now is to find a spot to sit and let other hunters drive the animals in my direction.

The prehistoric hunter undoubtedly went through similar experiences and likewise realized that there comes a time to turn hunting over to younger generations. I see at least three generations of younger hunters able to do what I was able to do some decades earlier; they are tolerant of but obviously tired of listening to my stories about the halcyon days of hunting. I hope they will eventually be able to reflect on their hunting experiences in the same way I do.

My years of experience with animals both wild and domestic and the use of weaponry both modern and prehistoric now add up to a total of nearly three-quarters of a century. The question that arises from the foregoing presentation is whether by drawing on my experience I can potentially produce better interpretations of the archaeological record. I believe the answer is a strong affirmative, but only when that experience is used in conjunction with pertinent data from many other disciplines. In this way, knowledge of animal behavior learned through actual subsistence hunting can be applied to the interpretation of archaeological assemblages; and the problems that inevitably arise from analyses of these assemblages can in turn be addressed by observing and experimenting with living animal populations. That is the approach used in the following pages as I discuss prehistoric large mammal hunters on the Great Plains and in the Rocky Mountains of North America. This book will not provide all the answers, but it should challenge other archaeologists to probe more deeply into the questions. I also believe that archaeologists and others can employ the kind of experimental approach described here, combined with data from archaeological site investigations, to better analyze and understand prehistoric strategies of animal procurement. Moreover, because different attitudes toward animals, new technologies, and changes in wildlife management brought on by increased human population pressures have rendered the old hunting philosophy and its practices obsolete with no prospect of revival, those practices must be recorded now—or never.

CHAPTER 3

Paleoindian Hunters and Extinct Animals

EVIDENCE FROM PALEONTOLOGY AND ARCHAEOLOGY

Ecological conditions on the Great Plains, and to a lesser extent in the Rocky Mountains to the west, were favorable for large grazing and browsing mammals; as a result, these two areas together became the main focus of land mammal hunting in North America. Earlier anthropologists wrote off the Great Plains as lacking any prehistory worthy of note before the introduction of the horse and the rise of the historic Plains Indian bison hunters (see Kroeber 1939; Wissler 1907). However, after the confirmation of human artifacts in association with extinct bison at the Folsom site in northern New Mexico (Figgins 1927), and with mammoths, also extinct, at the Dent site in northern Colorado (Figgins 1933) and at Blackwater Draw in New Mexico shortly afterward (Cotter 1937, 1938), doubts about the antiquity of humans in the New World faded, and several North American archaeologists turned their attention toward Paleoindian studies.

Acceptance of the fact that human hunters and extinct animals coexisted in North America gave rise to many pressing questions that required expertise from disciplines other than anthropology. These included questions about the taxonomy of the extinct animals; about the source, route, and time of human entry into the Americas; and about the ecological conditions encountered by these hunters on their arrival. Of equal interest was the reaction to them of the New World megafauna, which presumably had never before encountered human predators. The role of human predators, direct or indirect, in the demise of several late Pleistocene and early Holocene large mammal species is a topic that always generates lively discussion among both specialists and laypeople. Though some believe that human predation was a major factor in these extinctions (see Martin 1967), the issue remains unresolved (see Martin and Wright 1967; Martin and Klein 1984).

This area of research has steadily grown into a multidisciplinary effort in which many specialists in and outside of anthropology and archaeology are engaged. Some argue for climatic conditions that altered or destroyed the subsistence base; others believe the animals were unafraid of the newly arrived human predators, who rapidly drove them to extinction; still others take a middle ground and argue for both environmental deterioration and human predation (see L. Marshall 1984). It is also possible that some disease was introduced as humans entered the continent. But the archaeological record clearly demonstrates that human predators were present in North America during part of the extinction period, and their presence alone provides a legitimate reason for arguing that they contributed, in some way, to the animals' final demise.

I believe that the earliest New World human group known to have been present on the North American Great Plains with weaponry and tools capable of killing, butchering, and processing large mammals such as the mammoth and extinct bison was Clovis. Its first well-documented appearance in North America was between about 11,500 and 11,000 years ago and is identified by a distinctive type of weaponry produced by a specific lithic technology—that is, a very particular way of making stone weaponry. Moreover, Clovis is the only North American hunting group known to have been directly involved with mammoth procurement. The archaeological record has not yet revealed whether there were pre-Clovis groups on the Northern Plains and in the Rocky Mountains. However, no earlier cultural assemblage in North America possessed weaponry and tools that could inflict lethal wounds on large mammals and butcher and process the results, and thereby establish a viable subsistence strategy based on large mammal hunting (see Hofman 2000).

Clovis hunters were killing mammoths *(Mammuthus columbi)* and extinct species of bison *(Bison antiquus* and *Bison occidentalis)* more than 11,000 years ago. In addition, small amounts of pronghorn *(Antilocapra americana),* camel *(Camelops* sp.) and horse *(Equus* sp.) skeletal remains have been recovered from Clovis sites, suggesting they were also hunted but (given their limited numbers) contributed little to the total food supply. Clovis appears to have arrived on the North American Plains as a fully developed cultural complex from some other location, currently unknown. While the presence of these hunters in North America appears certain, there is much speculation concerning how and when they arrived. Four possible routes have been suggested: across the Bering Strait from Siberia and then south, following an ice-free corridor between the Laurentide and Cordilleran ice sheets; a Pacific coastal route from Siberia that avoided inland Alaska; a Pacific Ocean route into South America; and a North Atlantic Ocean route (see Dixon 2000). Unable to locate lithic assemblages that they believed could represent Clovis progenitors in northeast Asia, Dennis Stanford and Bruce Bradley (2000) have revived an old idea, suggesting that the unique similarities between Upper Paleolithic Solutrean and Clovis lithic technology could indicate a direct cultural connection. That the greatest concentration of Clovis projectile points in North America lies along the Atlantic coast strengthens this hypothesis. In any case, the argument for a Solutrean connection merits serious consideration but requires much more study and documentation before it can be confirmed or denied (see Straus 2000).

In addition to the large herbivores, several formidable carnivores became extinct during the waning years of the Pleistocene. They include the short-faced bear *(Arctodus simus)*, the American lion *(Felis atrox)*, the dire wolf *(Canis dirus)*, and the American cheetah *(Miracinonyx trumani)*. There is no confirmed human association with these four species, with one possible exception: a partial radius and a metacarpal of *Arctodus* from the Clovis level at the Lubbock Lake site in northwest Texas. The former bone may show evidence of use as a tool and the latter may have cut marks (Johnson 1989). That skeletal elements have been culturally modified reveals little if anything of the circumstances of the animal's death, however.

As a hunter, I would be extremely interested to know if Clovis hunters, at any time, came face-to-face with a short-faced bear. With its size, running ability, and dental array, it was undoubtedly a formidable carnivore; and if its temperament was analogous to that of the much smaller grizzly *(Ursos arctos)*, also present then and now, a hunter attempting to kill one would have faced real danger. I never was the actual target of a charge by a wounded grizzly, but more than one hunter has vividly described the experience to me, making clear that the bear—particularly a wounded one—can charge a person with frightening speed and ferocity. The novice grizzly hunter is well advised to take along an experienced hunter as backup.

The mammoth, horse, and camel apparently became extinct in North America by about 11,000 years ago, but the presence of extinct varieties of bison along with pronghorn in archaeological sites continues. Mountain sheep (*Ovis* sp.), mule deer *(Odocoileus hemionus),* and elk *(Cervus elaphus)* make their first known appearance in archaeological sites of about this age. There is evidence that the late Pleistocene mountain sheep was larger than the modern species and it has been assigned a different taxon, *Ovis catclawensis* (see Wang 1982). The absence of sheep, deer, and elk in archaeological sites earlier than 11,000 years old could be a sampling problem: this might change at any time as new evidence is recovered.

Natural Trap Cave (Gilbert and Martin 1984) in northern Wyoming and Little Box Elder Cave (Anderson 1968) in central Wyoming together produced much paleontological evidence of the late Pleistocene and early Holocene fauna, but neither assemblage provided evidence of human association (map 2). A short-faced bear was recovered in the Hot Springs, South Dakota, Mammoth site (Agenbroad 1990) but was of an age believed to predate any known human presence on the Great Plains or in the Rocky Mountains.

Stone weaponry and tools found in context with faunal materials in geologic deposits of high integrity usually satisfy most investigators as evidence of human involvement, and the projectile points are generally accepted as being instruments of some animal's demise. In the absence of such tools and weapons, it can be difficult to establish human association with faunal assemblages derived from other lines of evidence. It took some time before taphonomic analysis of paleontological and archaeological faunal assemblages became accepted; and even now, after it has reached a level of credibility, distinguishing among human, animal, and natural modification of skeletal material is straightforward in some cases and not so straightforward in others.

Because no one today can actually observe a prehistoric hunter kill a mammoth, camel, horse, or an extinct bison, and because the bone beds reveal very little if anything about the actual killing strategies, scholars often speculate about whether these hunting efforts reflect systematically planned procurement strategies, opportunistic encounters, or simply the scavenging of dead animals. It seems to me that the hunting as we are seeing it revealed in the bone beds might best be described as "systematically opportunistic." In any case, I strongly believe that only rarely, under unusual circumstances, would a reputable hunter scavenge dead animals.

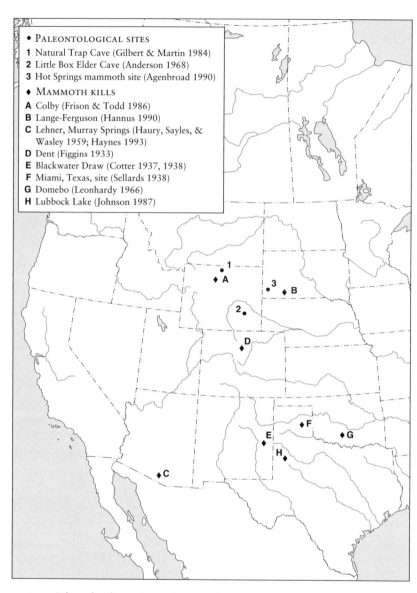

MAP 2. Selected paleontological sites and mammoth kill locations.

Some investigators have wondered whether the Clovis projectile points recovered in mammoth kill sites could have inflicted wounds lethal to an animal so large. These questions led to a number of different experiments on dead elephants (Huckell 1979, 1982; Stanford 1979b). Despite their many shortcomings, these tests were valuable in demonstrating the limitations of stone weaponry and tools.

Many species of animals that were hunted prehistorically are still present in their traditional habitats, thus allowing reliable behavioral studies. The same cannot be said for the extinct species of mammoths, bison, camels, and horses; nor do we know how closely the behaviors of these vanished animals would compare to that of their present-day equivalents—modern elephants, modern bison, modern camels, and modern feral horses. This large gap in our knowledge introduces an unwelcome element of uncertainty into our reconstructions of past procurement activities, which would be far more certain if we had more reliable ways of knowing the behavior of the animals involved. Yet the underlying principles of hunting are applicable even for extinct species.

CLOVIS HUNTERS

Clovis hunters possessed well-designed, well-made, and well-maintained weapons and tools, and developed systematic and effective large mammal procurement strategies. They did not depend solely on opportunistic encounters with prey species. The quality and complexity of the lithic, bone, and ivory components of known Clovis cultural assemblages indicate that these hunters may have lived in small, mobile social groups but certainly did not endure an impoverished, debilitating hand-to-mouth existence (see Butler 1963; Lahren and Bonnichsen 1974; Gramley 1993; Frison and Bradley 1999).

Although Clovis hunters may have hastened the demise of the mammoth, camel, and horse, it seems unlikely that they did so by circumventing the social restrictions found in today's primitive hunting societies. But because mammoths reproduced so slowly—the females bore young about every three years, beginning when they were about fifteen years old—the loss of breeding females to Clovis hunters could have dramatically hastened the extinction of the species, especially at a time when ecological conditions were deteriorating.

My introduction to mammoths and my speculations on mammoth hunting result from a purely serendipitous event in the Big Horn Basin of north-central Wyoming in the summer of 1962. A heavy equipment operator, Donald Colby, working on a highway project was moving dirt to enlarge a small stock-watering pond when he noticed something that proved to be a Clovis-type projectile point (figure 4b). Colby salvaged the artifact and showed it to Eugene Galloway, an archaeologist who happened to be in the area. Galloway sent me a picture of the point and a map of the location; two years later, I found several mammoth tooth plate fragments nearby. This discovery prompted annual checking of the same area that yielded nothing more until a heavy downpour in the late summer of 1973 exposed a concentration of mammoth tooth plate fragments from a badly deteriorated pair of mandibles, suggesting a possible connection with the Clovis point found eleven years earlier. Excavations at the site in 1973, 1975, and 1978 produced the partial remains of eight mammoths, one of which was a fetus. Three more Clovis projectile points recovered in context with the mammoth bones, assumed to be weapons used to kill the mammoths, leave little doubt of associated human activity. The site was named the Colby Mammoth Kill after its discoverer.

I was quite familiar with the location, but not in the context of a mammoth kill site. The reservoir being enlarged was originally used for livestock being trailed across a dry 40-kilometer stretch of dusty badlands between there and the Big Horn Mountains to the east. I had watered cattle at the pond numerous times, never suspecting the presence of mammoth bones. The small pond had trapped only runoff water, but an irrigation project nearby provided more water for the new reservoir. Had the pond not been enlarged, it is doubtful the site would ever have been found. A portion of the mammoth site is believed to lie still unexplored below the water level of the enlarged reservoir.

The bottom of a dry arroyo in an area of sparse vegetation and intense deepening of stream channels (downcutting) seemed an unlikely location for a Clovis-age mammoth kill to have avoided the destruction wrought by geological activity. However, geologic study of the Colby site area (Albanese 1986) has demonstrated that the mammoth remains lay in the channel of what was at the time a deep, steep-sided, intermittently flowing arroyo. Shortly after the bones were deposited, the geologic regime changed; colluvial and alluvial sediments accumulated above them to a depth of possibly as much as 10 meters. At some later date, the geologic processes reversed and downcutting began. Part of the present arroyo channel is at about the same depth as the old one but at some distance away from it, leaving some alluvial and colluvial channel deposits that contain part of the mammoth remains still intact in their original position. However, the area is now degrading, and without the chance

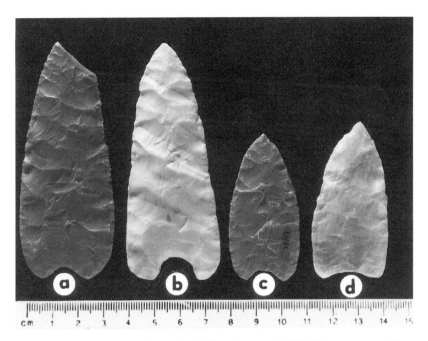

FIGURE 4. Clovis points from the Colby Mammoth Kill. Points *a, c,* and *d* were recovered during excavations; *b* is a surface find by Donald Colby. (From Frison & Todd 1986: 92.)

discovery of the Clovis point, the mammoth bones would almost certainly have weathered away unnoticed (they rapidly disintegrate to powder before they become visible at the surface).

Today's arroyo lacks the high, perpendicular walls of the one that existed at the time the mammoths were killed, whose original configuration and location suggest that it may have figured strongly in the hunters' procurement strategy. It was the only major topographic obstacle between there and the Big Horn River, a distance of nearly 7 kilometers. The hunters may have followed mammoth family groups from the river to the site, waiting for the opportune moment to launch a dart into an animal far enough away from the matriarch that her suspicions would not be aroused. At my suggestion and based on this hunting strategy, an artist, Roy Anderson, produced a painting depicting the event that was used in a *National Geographic* feature article (see Canby 1979: 346–47). Unfortunately, the figure caption does not relate the picture directly to the Colby site (and once again I must assert my reservations on the use of artists' renditions to describe prehistoric hunting). I do not think the evi-

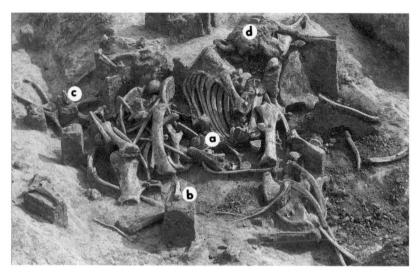

FIGURE 5. Mammoth bone pile no. 2 from the Colby site: *a*, discovery
location of the point (figure 4c); *b*, broken camel radius; *c*, bone tool made
from the femur of an unidentified medium-sized mammal; *d*, skull of an
immature male mammoth. (From Frison 1978: 96.)

dence from the Colby site points to a single kill involving a family of
mammoths. More likely the location favored efforts to finish off single
animals wounded on separate occasions. The presence of nursery herd
animals suggests that one or more experienced hunters may have followed
a family group and wounded young animals that were temporarily away
from the protection of a matriarch. If so, they could have been oppor-
tunistically attacking a mammoth family that systematically exploited a
territory much as African elephant families do. This assumes, of course,
that mammoths, in their use of a territory and their social structure, were
similar to the African elephants of today (see Laws, Parker, and Johnstone 1975).

    At the edge of the channel of the old arroyo were two concentrations
of mammoth bones. One was nearly intact and consisted of the articulated left scapula-humerus plus twelve ribs from the front quarter of a
mature female mammoth. Scapulae and pelvic and long bones from at
least three other mammoths were stacked around it (figure 5). The articulated section represents the only postcranial elements from a fully mature mammoth recovered at the site, although two skulls with mandibles
from mature females were present elsewhere in the old arroyo. Atop the
pile were the skull and mandibles of a juvenile male mammoth estimated

FIGURE 6. Mammoth bone pile no. 1 from the Colby site. The arrow indicates the point in figure 4a. (From Frison 1978: 94.)

to be five years of age. At the base of the articulated rib unit lay a Clovis point (figure 4c), a position strongly indicating that it was instrumental in the death of the animal.

The bone pile itself is controversial, because it appears to have been deliberately placed there. An Arctic anthropologist looking at pictures of the pile noted its resemblance to present-day frozen caribou meat caches: bones are stacked around the piles of meat, and the skulls placed on top mark their location in deep snow. The outsides of the piles are packed with slush, which freezes solid and helps protect them from carnivores and scavengers. No more plausible interpretation of the mammoth bone pile at the Colby site has yet been proposed. In today's climate, such a cache at this latitude would remain frozen during the cold months from November to March. If it was a cache, it was unopened: its contents would have spoiled once warm weather arrived. However, such wastage—commonly referred to as an "insurance cache" (see Binford 1981: 287)—was preferable to cached meat being needed but unavailable. Another feature a short distance away contains similar mammoth skeletal elements but in a more dispersed configuration (figure 6); it apparently was a meat cache that was used.

Part of the bone pile with the skull on top had been undercut by stream action that allowed some of the bones to move downslope toward the bottom of the arroyo channel. This circumstance prompted one investigator (Shipman 1981: 87–90) to suggest that the pile itself had resulted from stream action. To test that possibility, Lawrence Todd and I found a small stream with nearly the same gradient, sinuosity, and bed conditions; using bones from a mature female Asian elephant *(Elaphus maximus)*, we conducted a number of experiments on them by damming the stream and releasing different volumes of water. We found that stream flow tends to concentrate different bones at different distances and thus produce results different from those in the bone piles (see Todd and Frison 1986: 61–68). We believe that these experiments provide a strong argument against the two bone piles having been formed by stream action. Nevertheless, some of the bones in the channel most likely did assume their present positions as the result of intermittent stream activity. In one case, a skull with one tusk intact lay in the arroyo channel and formed a barrier that collected several other bones.

There are reliable eyewitness accounts of other mammoth bones found in 1909 in an arroyo adjacent to and a short distance from the Colby site. These were excavated during the building of a small earthen dam and were taken to a nearby town, where they were displayed for a short time; they rapidly deteriorated, however, and were discarded. There is no record of artifacts found with them, and the discovery location was later buried by highway construction. This earlier find further supports the contention that the Colby area was utilized for mammoth procurement over a period of time and that the bones at the site are not evidence of a single event.

Other Clovis artifact and mammoth associations on the Northern Plains (see map 2) include the Dent site along the South Platte River in eastern Colorado (Figgins 1933) and the Lange-Ferguson site east of the Black Hills in western South Dakota (Hannus 1990), with twelve mammoths at the former site and two at the latter. Clovis mammoth kills on the Southern Plains include Blackwater Draw (Figgins 1933), Miami (Sellards 1938), Domebo (Leonhardy 1966), and Lubbock Lake (Johnson 1989). Lehner (Haury, Sayles, and Wasley 1959) and Murray Springs (Haynes 1993) are in southeast Arizona (see map 2). Possible but questionable human-mammoth associations were found with the Lindsay site in eastern Montana (Davis 1971), the Union Pacific Mammoth site in southern Wyoming (Irwin and Agogino 1962), and the Rawhide Butte Mammoth (Damon, Haynes, and Long 1964) and Jewett Mammoth sites

(Zeimens 2001), both in southeast Wyoming. No human-mastodon associations have been confirmed in this area.

OTHER EXTINCT MAMMALS

Included in the Colby site faunal assemblage was the humerus of a perissodactyl *(Equus conversidens)*, but with no convincing evidence that it had been killed by Clovis hunters. One of the more vexing problems posed by the late Pleistocene extinctions is the difficulty of explaining why bison were able to survive and horses were not. The question is particularly puzzling given that feral horses, probably descendants of animals that escaped from the herds of both Native Americans and early settlers, have flourished in remote and inaccessible parts of the Big Horn Basin and southwestern Wyoming for more than a century. After a generation or so in the wild, they rapidly revert to a nondomesticated pattern of behavior. Although protected today, they were not well regarded during the early part of the twentieth century: they multiplied rapidly and they were difficult to capture and train as either draft or riding stock. Their ability to chew and to paw for food in deep snow made them a threat to the grasslands. They also threatened livestock: wild stud horses would regularly abscond with mares from domestic herds, and feral mares often attracted domestic stud horses. I have observed feral horses stampeding cattle around water holes and sometimes crippling or killing one or more of the herd, behavior that did not endear them to ranchers. In addition, feral studs could be mean, especially if a mare was in heat, and would sometimes attack a person on foot or horseback. As the numbers of feral horses grew rapidly in the early decades of the twentieth century in the absence of natural predators, about the only means of control was to shoot them on sight. They were killed strictly to limit their population: I knew of no one who was reputed to have salvaged horse meat for human consumption.

One of the most pitiful sights I remember from my early youth was an old feral mare whose young foal had died from lack of milk. The badlands of the Big Horn Basin have many areas of bentonite clay, which is a sticky gumbo when wet and cement-hard when dry. It was late March, and as the mare's tail dragged along the ground during the day it had picked up the sticky clay that then froze solid at night, gradually forming a wad about a foot in diameter. The mare managed to struggle to her feet and made feeble efforts to run away from us with the ball of clay swinging from side to side, hitting her in the ribs first on one side and then the other. In addition, her long mane and foretop were a solid mass of cockle burrs; her neck and lower jaw were covered with dozens of Rocky Mountain wood ticks, many expanded to nearly a centimeter in diameter with the old mare's blood; and her front teeth were so worn away that she could no longer ingest enough grass to survive. Her lack of mobility had allowed her hooves to grow out, further restricting her movement severely. One of our group mercifully ended her misery with a bullet to the brain. Two decades earlier, a horse in such weakened condition would undoubtedly have been pulled down by wolves earlier in the winter.

To date, the lack of any satisfactory evidence on the role of the Pleistocene horse in the hunting efforts of Clovis groups makes any ideas on their procurement pure speculation. Any comparison of feral and Pleistocene horse behavior is also speculative. However, it would be relatively easy for hunters using Clovis weaponry to devise a strategy to procure feral horses. A stud horse defending his mares is a bit like a matriarch elephant defending her family. The biggest difference is that the stud has to continually fight off other males seeking to take over his harem, while a matriarch elephant's status as the head and protector of her family is undisputed.

The case for Clovis-age horse procurement may have been strengthened by recent discoveries at the Wally's Beach site located in southern Alberta. The lowering of the water level in a shallow artificial lake exposed loosely consolidated sediments; as they were removed by high winds, trackways of extinct mammals—including mammoths, camels, and horses—were revealed. Two broken Clovis projectile points recovered as surface finds in the area yielded horse-protein residue, suggesting that they had been used either as weapons to kill horses or as tools to butcher them (Kooyman, Newman, et al. 2001). Along with the animal tracks were bones of horse and bison. The radiocarbon date of the horse bone collagen is 11,330 B.P. ± 70 years (TO-796). The investigators believe that the bone distribution and the presence of stone tools are strong evidence that the extinct horse *(Equus conversidens)* was butchered (Kooyman, McNeil, et al. 2002).

Even before feral horses were protected by law, they managed to survive in large numbers in unfenced areas of rough terrain. Some systematic trapping was done using strategically placed corrals with long, expanding wings reminiscent of historic bison corrals. Before airplanes and helicopters existed, experienced riders with high-quality mounts were needed if feral horses were to be corralled with any degree of success.

Chasing them actually became a form of entertainment, though good horses were sometimes ruined in the process; one or more riders could often lasso and capture a feral horse. At least one mustanger, as those in the business of capturing wild horses were known, was alleged to deliberately release a well-bred stud into the feral herds—an easy way to upgrade the animals and increase the profits from the animals he would eventually capture.

I believe there is a valid reason to explore feral horse behavior in the context of historic Native American hunting. Horses are unquestionably central to today's popular stereotype of the Plains Indian, and there has been much speculation among academics about when and under what conditions Native Americans acquired horses and about their influence on Plains Indian warfare and bison hunting (see, among many others, Wissler 1914; Haines 1938a, 1938b; Secoy 1953; Denhardt 1947; Ewers 1955). Once the potential of horses was realized, there was a rapid shift from hunting on foot to pursuing animals on horseback. The mounted hunters were able to choose among more bison procurement strategies, significantly increasing both their flexibility and their effectiveness. In fact, some of the old pedestrian methods of bison hunting, bison jumping in particular, were wholly abandoned in favor of pursuit of bison on horseback.

Horses were introduced to the New World by the Spaniards, and Native Americans obtained them by raiding Spanish herds and capturing feral animals. Much of New Mexico, parts of northern Mexico, and the coastal part of Texas offered favorable horse habitat (Secoy 1953: 3) and became populated with feral herds soon after the earliest Spanish explorers arrived. Native Americans first recognized horses as a source of food but soon came to appreciate their value in hunting and warfare. John Ewers's informants claimed that herds of feral horses were present in Blackfoot territory but that they had little success in catching the animals, many of which died afterward (Ewers 1955: 59). On the other hand, some Southern Plains tribes were said to have become adept at capturing feral horses using a lasso and a running loop, or an open loop on the end of a forked stick (Ewers 1955: 60).

In my own limited experience, when animals as greatly desired as horses are roaming free and available for the taking, white men do not pass up the opportunity offered; I doubt that Plains Indians would have reacted much differently. Yet more is involved than just locating a herd, slipping a hackamore on a feral horse, and leading it home. Even after one is caught, much time and effort are needed to mold it into a useful animal. For that reason, notes Robert Denhardt, a man whose writing reveals his wide knowledge and deep understanding of horses, "the natives [Southern Plains tribes] obtained their original horses, and always by far the greatest number, from the Spaniards or neighboring tribes and not from the wild herds. The Indians had mounts by the time the wild herds dotted the plains, and they always preferred domesticated animals to the *mestenos*. Mustangs were hard to catch and, once caught, harder to tame" (Denhardt 1947: 104). However, "the greatest number" is not "all," and it is likely that the Southern Plains tribes captured some feral horses.

The Plains Indian ethnographic accounts emphasize the value placed on horses used for chasing and killing buffalo. In my own experiences with horses and hunting, I have found that some horses, like some humans, become addicted to pursuing large wild animals. My favorite hunting horse underwent a complete change in attitude, transforming from a plain old cow horse to a hunting horse, whenever I tied a rifle and scabbard on his saddle and headed for the mountains. He would nearly always spot an elk herd before I could, and then champ at the bit to be allowed to begin the chase. I believe that the ethnographers give too much credit to the Plains Indian bison hunter for training a horse and not enough to the occasional horse that developed a lust for chasing bison. Descriptions of eighteenth-century vaqueros in California who hunted grizzlies on horseback, using rawhide reatas to immobilize and strangle them, support this surmise about horse behavior. Several accounts agree that "perhaps the most remarkable feature of the struggle was the sagacity and skill of the horses and their own delight in mastery of the bear" (Storer and Tevis 1978: 134).

CAMELS

I can claim no experience with camels beyond having ridden one on a few occasions. Their gait, temperament, and general behavior are entirely different from those of a horse. The artist's conception of camel hunting in the 1967 edition of *Pleistocene Extinctions* (Martin and Wright 1967: 253; see also Wormington and Ellis 1967: frontispiece) is unconvincing as a portrayal of a believable hunting strategy. The camel has the end of a spear shaft protruding from one side of its rib cage and apparently another on the opposite side. The wielder of one spear has carelessly allowed himself to approach too close to the animal's front foot, a location where he could easily suffer a broken leg. Another hunter is approaching the animal from the rear with a stick apparently intended for use as a club; a second seems to be picking up a rock, presumably to throw at the animal; a third is shaking a branch at the animal's head; and yet a fourth is preparing to throw another spear, apparently at the head or neck. In reality all they need to do is withdraw and allow the effects of the two chest wounds to weaken the animal. A juvenile camel, probably the offspring of the first one, has also been speared in the rib cage and is unnecessarily being clubbed into final submission by two other hunters wielding oversized tree branches. The illustration is a fanciful and unrealistic portrayal of hunting and killing animals.

Occasionally, small amounts of camel remains appear in early Paleoindian archaeological sites; lacking adequate comparative skeletal material of *Camelops,* I was able, several years ago, to exhume several complete modern camel skeletons that had died at a wildlife farm near Tehachapi, California. The modern skeletons are useful, because the modern animals are quite close in size and morphology to their prehistoric relatives. Those camel remains that have been found in archaeological sites on the Great Plains provide little solid evidence of having been killed and eaten by Paleoindians. I first believed that the partial remains of a camel recovered in the 10,000-year-old bison bone bed at the Casper bison kill site were of Hell Gap age (Frison 1997: 12–16), but the radiocarbon date of the camel's calcaneus is 11,190 B.P. ± 50 years (CAMS-61899), more than a thousand years older than the bison bones (Frison 2000a). Though one humerus is broken in a manner strongly indicative of a green bone fracture, we cannot conclude that the break was the result of human activity.

HUNTING AFRICAN ELEPHANTS

The question of whether a Clovis projectile mounted on the end of a thrusting spear or thrown with an atlatl could deliver a lethal wound to an animal the size of a mammoth has been raised numerous times. Several experimental sessions, using dead elephants from zoos and game farms (see Huckell 1979, 1982; Stanford 1979b; Weaver 1985), demonstrated that Clovis points would penetrate elephant hide but could provide little information about possible strategies for hunting elephants. I have always strongly advocated realistic experiments, and I hoped someday to observe elephants in their natural habitat and to gain enough experience to propose a procurement strategy for African elephants that might also apply to mammoths.

I was able to measure the thickness of several preserved pieces of mammoth hide in the collections of the Zoological Museum in St. Petersburg, Russia; I found that it was similar to the hides of both African and Asian elephants, but I could not determine whether the mammoth's hair would more seriously impede a Clovis projectile point than does a bison's winter coat. For several years, I sought some way to experiment with Clovis weaponry on modern elephants. Finally, in the mid-1980s I had an unusual opportunity that offered a close approximation to hunting live African elephants.

Because elephant numbers in parts of Zimbabwe were too high, causing serious loss of habitat, culling operations were initiated. Gary Haynes from the University of Nevada, Reno, who had established a long-term research project with elephants in Zimbabwe, was able to convince wildlife managers in Hwange National Park in Zimbabwe to allow me to experiment with Clovis weaponry on the dead and dying elephants that had been culled. Bruce Bradley made several Clovis point replicas from different lithic materials (figure 7). In 1984, my experiments were with thrusting spears; in 1985, I used an atlatl and dart, copying the wooden and other perishable components from Archaic-age specimens found in dry caves in the Big Horn Mountains of northern Wyoming (Frison 1965, 1968a). I did not attempt to use a throwing spear. The experiments were useful, allowing me not only to demonstrate the use of stone weapons and tools in killing and butchering elephants, but also to closely observe African elephant behavior (Frison 1989).

Culling elephants is a textbook example of why one needs to understand animal behavior to successfully carry out such hunting ventures. Indiscriminately killing the elephants proved to be the wrong approach to reducing their numbers. Elephant families are structured with a matriarch as the leader and protector of her offspring, and it was necessary to take the whole nursery herd, which contained as few as eight to as many as fifty individuals. Family members that evaded the culling would not be accepted by other elephant families and would become a threat to visitors to the national park. Another reason for pursuing this strategy was the long memory of matriarch elephants: if one connected a human hunter with harm previously done to one of her offspring, that person could always be in some danger if ever in her vicinity.

The best strategy for eliminating an entire family is to first kill the matriarch; as the rest of the family crowds around her body seeking protection, they are quickly and easily dispatched. The matriarch is prodded into charging the shooter, who drops her in her tracks with a

FIGURE 7. Experimental Clovis point mounted in a split foreshaft and inserted into a mainshaft. (From Frison 1989: 772.)

large-caliber, jacketed bullet to the brain, which lies well toward the back of the skull (figure 8). The bullet must pierce hide and bone for some distance: a soft-nosed bullet, which expands and quickly loses velocity after impact, would not be able to penetrate the distance necessary to enter the elephant's brain. At first, the shooter approaches the elephant family close enough that the matriarch perceives a threat. She will make a false charge or two, giving the shooter a chance to withdraw before she makes a final charge. It requires a high level of courage, know-how, and confidence in one's weaponry to face a charging matriarch. The shooters I observed never failed to drop the animal with the first shot. They knew when she was making the final charge, they knew the location of her brain cavity, and they were good enough marksmen to place a bullet in the exact spot needed. It is not an activity for the inexperienced hunter or the faint of heart.

I learned a great deal from participating in the elephant culls. Mam-

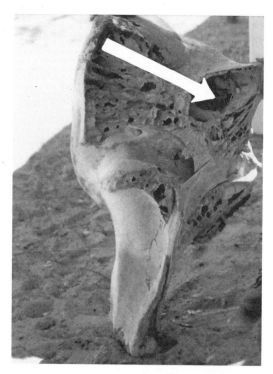

FIGURE 8. Split African elephant skull. The
arrow indicates the brain cavity. (From Frison
1991b: 148.)

moth anatomy is similar enough to that of African elephants to make it
obvious that a hunter would be very ill-advised to try to kill either animal by attempting to penetrate its brain cavity with a Clovis projectile,
whether delivered by a thrusting spear or by atlatl and dart. There is simply too much hide and bone protecting the brain. The throat is a small
target: the chance of penetrating it with a stone projectile point at the
right location to sever the main vein and cause the animal to bleed to
death is quite small. In addition, the attempt often results in the projectile lodging in one of the cervical vertebra. Any shot toward the vicinity
of the head of an African elephant can be deflected by their large, continually moving ears; such deflection would have been less of a worry
for those hunting mammoths, because their ears were smaller.

    The heart of an African elephant is another poor target for Clovis
weaponry, since it lies close to the bottom of the rib cage and is well protected by the flattened and widened distal ends of the ribs. A stone pro-

jectile point cannot penetrate an elephant's rib; if it hits the distal, flattened end, it will usually break or shatter on impact. Further up toward the midsection, the rib is rounded; a projectile point will usually slide around it and enter the chest cavity, though it may snap in two in the process. Yet even if broken, it may continue into the animal and be able to produce a lethal wound.

Intestinal and flesh wounds should be avoided if possible, despite their potential (depending on their location and depth of penetration) for eventually proving to be fatal. Intestinal wounds may require several days to bring about the death of an animal. Any serious wound will usually weaken the animal quickly, causing it to go into a crouching position or lie down. The best strategy at such a time is to leave the animal alone and allow the debilitating effects of the wound to continue. After an appropriate wait, a hunter may then carefully approach and deliver the lethal blow. However, if a wounded animal is approached too soon and is able to get to its feet, it may travel a long distance before lying down again. Patience and caution usually pay off when one is pursuing injured animals.

To kill an elephant with atlatl and dart, the optimum strategy is to be patient and wait for a broadside penetration of the upper rib cage into the lung area. Once the point is inside the lung cavity, any movement of the animal causes internal hemorrhaging. Being able to select and deliver the weapon to the best location is crucial: too far forward, and the flesh and bone of the scapula lying on the rib cage block the projectile; too far to the rear, and an intestinal wound results. The cooperation of two or even three persons with long experience hunting together will increase the chance of success. Young elephants in particular often lose immediate contact with the rest of the family and the protection of the matriarch, leaving them vulnerable to a human predator. Elephants have relatively poor eyesight but an acute sense of smell. The experienced elephant hunter carries a cloth bag filled with fine dust that he can shake to test the direction of air movements and ensure that he stays downwind. Once again, we need to remember that mammoth behavior may have differed significantly from African elephant behavior.

These African elephant experiments highlighted a number of concerns about the use of primitive weaponry in killing large animals with thick hides. Success requires a very strong grip and considerable force on the shaft of a thrusting spear to penetrate the thick, tough hide of a mature elephant. The projectile point needs to be of good-quality stone with a sharp tip so that it can penetrate and not shatter on impact. Blade edges need to be sharp so that they can cut a hole large enough for the fore-shaft to enter. A relatively light sinew binding proved adequate as long as the projectile was carefully fit into the nock of the foreshaft; too much binding hampered the entry of the blade and shaft into the animal. At its point of entry, the projectile must be as close to perpendicular to the animal as possible.

Hidden flaws such as crystal pockets and internal fractures in the lithic material may not appear until it is placed under extreme stress; they usually result in a break that renders the stone projectile point totally ineffective. Such a break generally ruins a hunting effort; and when dangerous animals are the target, it can also lead to the hunter's injury or death. To avoid these serious consequences, weaponry needs to be carefully inspected and tested.

Wooden parts of the weaponry assemblage also have to withstand severe stress, and their failure is not uncommon. The taper of the socketing end of a foreshaft must match closely the taper of the hole in the mainshaft. Foreshaft, mainshaft, and projectile point must be perfectly aligned so that the force of the thrust is applied to the center line of all three components. Any small deviation can and usually does result in breakage, failure to penetrate, or both. Because wooden parts tend to warp and twist even if well cured, proper storage is critical. Also, woods vary in quality. Willow proved unable to penetrate elephant hide, but it is adequate for use on smaller animals, such as deer and pronghorn, with thinner hides. Shafts of chokecherry were stronger but more susceptible to warping. On one occasion, a shaft carelessly left leaning against a tree overnight warped enough to render it nearly useless. On yet another occasion, improper contact between the atlatl spur and the cup on the proximal end of the mainshaft prevented the latter from flying true and penetrating the animal.

Yet I was agreeably surprised at the effectiveness of the atlatl and dart in cutting through elephant hide and producing lethal wounds. Unfeathered shafts at a distance of 20 meters repeatedly entered the rib cages of elephants of nursery herd animals of all ages (figure 9) as well as one large mature male that unexpectedly slipped into a family group one night in response to a female in heat. Separate experiments using targets indicate that feathered main shafts improve accuracy, especially at distances beyond 20 meters.

I make no claim to have duplicated a strategy of Clovis mammoth procurement. These experiments were undertaken to demonstrate that a Clovis point with a simple delivery system such as a thrusting spear or

FIGURE 9. Experimental mainshaft with point and foreshaft in figure 7 buried in the rib cage of a mature female African elephant. (From Frison 1989: 777.)

atlatl and dart can penetrate an African elephant and produce a lethal wound. Nor do I claim that these items—with the exception of the projectile points manufactured by Bruce Bradley—replicate Clovis weaponry assemblages. The atlatl and shaft designs were based on 2,000-year-old Late Archaic weapons found in dry caves.

The skinning, butchering, and processing that have to follow the demise of an elephant to use its products and prevent spoilage can pose a challenge even if the animal is only partly grown. As noted above, the thickness of African elephant hide is comparable to that of mammoths: a sharp-edged stone tool, wielded with considerable pressure, is needed to cut through elephant hide, and the same undoubtedly held true for a mammoth. Not all tool stone is equally effective in cutting hide. My own preference is for a tool of quartzite, which maintains a sharp edge better than one made of chert—perhaps because the grainy composition of the quartzite allows the cutting edge to wear unevenly, while the edge of a chert tool dulls evenly and rapidly loses its ability to cut. One thing I soon learned in skinning large animals with stone tools is that once a tool loses its keen edge, there is no alternative but to resharpen it. Yet a tool edge too dull to cut hide will still cut flesh and sinew. Tool manu-

FIGURE 10. Cut in the hide of a freshly killed
juvenile male African elephant made with the tool
shown to the right of the cut. The tool is 81 mm
long and 40 mm wide. (From Frison 1989: 780.)

facture, sharpening, and use on animals of different sizes will be examined further in chapter 8.

I had the most success in butchering an elephant with a large, relatively thin quartzite flake, triangular or trapezoidal in transverse cross section, which can be continually resharpened laterally on the back side while leaving the bulbar side unmodified. Choosing a juvenile male elephant with hide that averaged 8.7 millimeters in thickness, I made a cut 25 centimeters long with such a tool before its edge became too dull for me to continue (figure 10). The cutting edge was slightly convex, 52 millimeters in length, and maintained at 45 to 47 degrees. The cut was made from the outside hide surface; however, once a cut is long enough that the skinner can gain access to and cut from the flesh side, the tool will not dull as rapidly, probably because it no longer comes into contact with abrasive material on the outside surface of the hide. A chert flake of similar configuration dulled sooner than the quartzite one.

The earliest known successful large mammal hunters in North America were those now called Clovis. They developed a weaponry system capable of killing mammoths and, to date, represent the only North American group unequivocally known to have hunted them. Evidence of other large and medium-sized mammals, including camel, horse, bison, and pronghorn, are found in Clovis sites but in relatively small numbers. There is no convincing evidence of Clovis hunters actually killing large extinct carnivores such as the short-faced bear, American lion, dire wolf, and American cheetah. Clovis hunters may have hastened the extinction of mammoths, but their role in bringing about the extinction of much of the late Pleistocene–early Holocene fauna is open to debate.

The true nature of Clovis mammoth hunting is also debatable; some think it opportunistic while others believe it to have been systematic. The procurement strategy at the Colby Mammoth Kill in northern Wyoming may be seen as falling somewhere between those two possibilities. This site also contains evidence of what may have been short-term, cold-weather caches of mammoth meat. Frozen meat caches may have been an important strategy for short-term meat preservation in other contexts (see Frison 1982d). For example, a pile of articulated bison skeletal units with individual bones scattered around it at the Agate Basin site is believed to have been a frozen meat cache from which meat was withdrawn as needed by the hunting group camped nearby (Frison and Stanford 1982b: 363). And for extinct North American species that possess modern relatives with similar skeletal elements—horses, camels, and mammoths—it is possible to study the modern animals to plausibly speculate about behavior patterns and procurement strategies.

CHAPTER 4

The North American Bison

The earliest North American bison derived from a species that originated in the Old World and migrated across the Bering Strait into North America sometime during the Illinoian glacial period of the Pleistocene, which began about 0.5 million years ago and ended about 0.125 million years ago. Leg bones were larger in all proportions, indicating a bigger, heavier animal standing higher off the ground than modern bison. Most noticeable are the larger skull and the widespread horn cores; earlier taxonomists relied on them to differentiate between a number of species, unaware that these were the most variable features of the entire skeleton. Because many of the early specimens were represented by little more than a partial skull and horn core, the result was a bewildering array of species designations (see M. Wilson 1975).

A consensus has not been reached by all investigators on the identity of the Asian species ancestral to the North American bison, nor on the details of bison evolution (see McDonald 1981). The first bison native to North America is referred to as *Bison latifrons,* and it is the largest North American fossil bison (figure 11). As more evidence accumulated, and as taxonomic trends changed, the late Pleistocene and early Holocene bison began to fall into two species designations, *Bison antiquus* and *Bison occidentalis.* The present-day North American animals (commonly

FIGURE 11. Mr. and Mrs. Jerry Peery of Canadian, Texas, with *Bison latifrons* horns from Lipscomb County, northeast Texas. (From Wyckoff & Dalquest 1997: 12. Photo courtesy of Don Wyckoff.)

known as "buffalo") are usually referred to as the plains-dwelling bison *(Bison bison bison)* and the northern wood bison *(Bison bison athabascae)*, which are difficult if not impossible to distinguish based on skeletal material alone. However, Jerry McDonald (1981) prefers to see the late Pleistocene–early Holocene bison as two subspecies, *B. antiquus antiquus* and *B. antiquus occidentalis,* and labels today's animals *B. bison bison* and *B. bison athabascae.* And Michael Wilson (1975), who made an exhaustive study of all bison then known from Wyoming, agrees with McDonald on the designation of modern bison but refers to the earlier ones as *B. bison antiquus* and *B. bison occidentalis.* Bison taxonomy is a topic that always generates lively discussion but has little relevance to prehistoric hunting. However the animals are labeled, the archaeological record reveals many similar strategies for bison hunting over more than 11,000 years, and, based on these similarities, I am assuming that the behavioral patterns of the extinct species resembled those of the modern species.

Though postcranial skeletal elements of extinct bison are larger than those of the modern species, the investigator needs to exercise caution in using size to make identifications, because measurements of bones of female extinct bison can be very close to those of male modern bison.

In the past, this fact has resulted in errors in judgment at some kill sites. For example, the bison from the first excavations at the Agate Basin site in eastern Wyoming were discarded because they were believed to be modern (Roberts 1943, 1961). This mistake was particularly unfortunate because the large bison bone bed in the area excavated (see Frison 1982c: 12) would undoubtedly have been a rich source of statistical data on Agate Basin–age animals of about 10,000 years ago. When skulls and horn cores are available, variations in their size and shape are the preferred means of distinguishing the extinct forms from the modern ones (figure 12). However, successful determinations require a valid sample of a bison population for analysis, because any individual animal might be at either extreme of the size range. Though numerous paleontologists and archaeologists have discovered fossil bison on the Southern Plains, Don Wyckoff and Walter Dalquest (1997: 21) are forced to admit that "we aren't even sure of differences between bulls and cows of the late Pleistocene large-horned bison forms." Clearly, much remains to be learned about fossil bison over the entire Great Plains.

According to one strain of earlier thought, the extinct bison represented evolutionary dead ends and the modern species resulted from a separate and later migration from Asia. However, it is now generally accepted that *B. latifrons* was the ancestor of both *B. antiquus* and *B. occidentalis,* which in turn were ancestors of the modern bison—present, according to the archaeological record, by about 5,000 years ago. *B. bison bison* most likely shrank to their current size in response to gradually changing climatic conditions that altered the ecology of the plains, foothills, and mountains. These changes may have begun as early as 11,000 years ago, at the end of Clovis times. Shortly after 8,000 years ago, a dry climatic period, commonly referred to as the Altithermal, began; it lasted for about 3,000 years. The archaeological record of this time reveals decreased human utilization of the plains, as well as few if any bison except in the foothills and mountains and other locations with more favorable conditions toward the north and northeast borders of the plains. A limited but sound body of archaeological site data confirming that these animals fell in size between the larger early Holocene bison and the smaller present-day bison is used to present a case for a gradual diminution of bison size from Paleoindian times to the end of the Altithermal period around 5,000 years ago. At that time, *B. bison bison* reappeared on the plains and were continually hunted until they were nearly extinct in the late nineteenth century. In archaeological sites containing bison over a large area of western North America, we find evi-

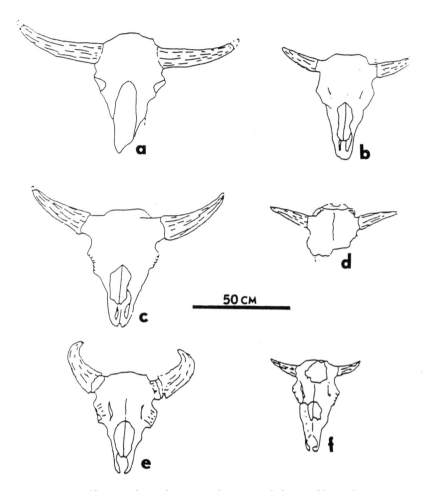

FIGURE 12. Changes through time in the size and shape of bison horn cores. Male *(a)* and female *(b) Bison antiquus* from the 10,000-year-old Casper site; male *(c)* and female *(d) Bison occidentalis* from the 6,500-year-old Hawken site; male *Bison bison (e),* with horn sheaths, that died in the late nineteenth century; female *Bison bison (f)* from the 300-year-old Glenrock Buffalo Jump. (From Frison 1998: 14578.)

dence of different procurement strategies used from the late Pleistocene until historic times (map 3).

Bone beds created both by human kills and by natural events (e.g., winter kills that resulted in large concentrations of animals dying at a single time) provide the best samples for population studies of prehistoric bison. Fortunately for archaeologists, some of the locations chosen

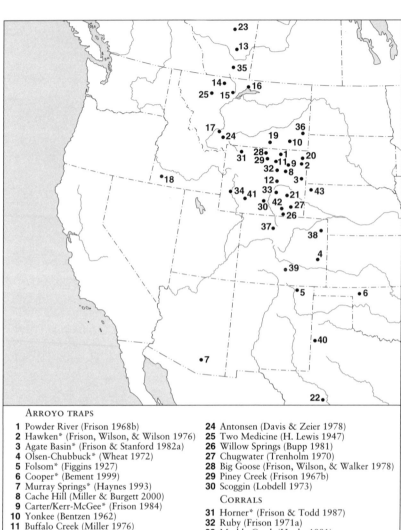

Arroyo traps

1 Powder River (Frison 1968b)
2 Hawken* (Frison, Wilson, & Wilson 1976)
3 Agate Basin* (Frison & Stanford 1982a)
4 Olsen-Chubbuck* (Wheat 1972)
5 Folsom* (Figgins 1927)
6 Cooper* (Bement 1999)
7 Murray Springs* (Haynes 1993)
8 Cache Hill (Miller & Burgett 2000)
9 Carter/Kerr-McGee* (Frison 1984)
10 Yonkee (Bentzen 1962)
11 Buffalo Creek (Miller 1976)

Sand dune trap

12 Casper* (Frison 1974)

Bison jumps

13 Head-Smashed-In (Reeves 1978b)
14 Boarding School (Kehoe 1967)
15 Ulm Pishkun (Davis 1978)
16 Wahpa Chu'gn (Davis & Stallcop 1966)
17 Logan (Malouf 1962a)
18 Five Fingers (Agenbroad 1978a)
19 Kobold (Frison 1970b)
20 Vore (Reher & Frison 1980)
21 Glenrock (Frison 1970a)
22 Bonfire* (Dibble & Lorrain 1968)
23 Old Women's (Forbis 1962a)

24 Antonsen (Davis & Zeier 1978)
25 Two Medicine (H. Lewis 1947)
26 Willow Springs (Bupp 1981)
27 Chugwater (Trenholm 1970)
28 Big Goose (Frison, Wilson, & Walker 1978)
29 Piney Creek (Frison 1967b)
30 Scoggin (Lobdell 1973)

Corrals

31 Horner* (Frison & Todd 1987)
32 Ruby (Frison 1971a)
33 Muddy Creek (Hughs 1981)
34 Wardell (Frison 1973)

Uncertain kill strategy

35 Fletcher* (Forbis 1968)
36 Mill Iron* (Frison 1996)
37 Upper Twin Mountain* (Kornfeld, Frison, et al. 1999)
38 Jones-Miller* (Stanford 1978)
39 Cattle Guard* (Jodry & Stanford 1992)
40 Plainview* (Sellards, Evans, & Meade 1947)
41 Finley* (Moss et al. 1951)
42 James Allen* (Mulloy 1959)
43 Hudson-Meng* (Agenbroad 1978b)

*Denotes fossil bison

MAP 3. Selected bison kill site locations.

by humans for communal kills were well suited for preserving bones. By borrowing analytical principles of site formation processes derived from geology, techniques of taphonomy from paleontology, knowledge of animal behavior from biology, and principles of human behavior from anthropology, researchers have gradually developed a useful methodology for studying prehistoric animal populations as they are found in archaeological bone beds and pertain to human hunting and its related social and ritual activities.

Because human involvement in bone bed formation is not always straightforward, the field of study is still evolving. Investigators rely heavily on analysis of bones to determine whether the causes of modifications are human, animal, or natural, and their conclusions are not always unequivocal. Numerous factors, including the length of exposure before burial, soil conditions, and bioturbation after burial, can cause bones to deteriorate and thereby erase evidence. Yet certain bone modifications are nearly always unique to human activities of hunting, butchering, and processing. A projectile point lodged in a bone convinces nearly everyone of a past human hunting effort; and it is fairly easy to tell when a bone has been broken on an anvil with a hammer stone rather than chewed by carnivores or scavengers or trampled by large animals. Green bones from freshly killed animals fracture differently than those that have been exposed to the elements for even a short period of time. A wide range of cut marks, impact fractures, abrasions, and other marks resulting from human tool use can usually be distinguished from nonhuman activities, but there are also many gray areas. Researchers soon learn to be cautious, because mistakes in identification result in misinterpretations that are seldom forgiven or forgotten by other investigators and invariably come back to haunt the individual responsible.

Still, we can claim with confidence that with the possible exception of the mammoth in Clovis times, bison were the major prehistoric human food source on the plains of North America, beginning with the earliest humans there. The discovery of the Folsom site in northern New Mexico (Figgins 1927) marks a major event in the study of both bison and human occupation of the plains. The bison were several animals of an extinct species killed at one time at a single location, and the associated weaponry indicated a group of sophisticated pedestrian hunters present more than 10,000 years ago. The relationships of the Folsom site hunters or any of the Plains Paleoindian large mammal hunters to those in the Old World are unknown. In the chronological sequence of projectile points from Clovis to Frederick (Irwin-Williams et al. 1973), we see more than 2,000 years of occupation by many different large-mammal-hunting groups with well-designed weaponry and no proven close counterparts in the Old World. It is possible, however, that these technological innovations might have been contributed by new hunting groups arriving in North America with their own long traditions of experience in hunting large mammals.

BISON BONE BEDS

Prehistoric bison hunters on the Great Plains used numerous procurement strategies intended to kill several animals at a single location at one time. These communal hunts, because of their size and complexity, are the most visible archaeologically and their success depended on the coordinated efforts of groups of varying sizes. Yet such hunts probably accounted for only a small percentage of the total number of bison killed by prehistoric hunters.

Landforms were important: in some cases a natural feature needed little if any modification to trap and hold animals. Artificial structures such as corrals or pounds were labor-intensive to build and maintain, particularly in the absence of what we view as the most basic metal tools for cutting trees and digging postholes. My many years of experience with bison confirm that to construct a corral that will hold them requires sizable posts and logs: bison are large, powerful animals, and when excited they can destroy an enclosure that has been carelessly constructed. Posts solidly anchored into holes make the strongest corrals with the least amount of building materials; but "buck fences"—large logs, well braced with posts above ground—are an alternative in areas where postholes are impractical or impossible. A corral implies wings that extend for varying distances to funnel the animals into its narrow entrance.

The only reliable test for any kind of natural bison trap or artificial corral is practical: bison may refuse to enter a corral or an arroyo trap or balk at approaching a precipice that to the hunters involved appeared entirely suitable. Sometimes minor changes in the direction of approach or the position of the wings will remedy its failings; sometimes the only solution is to abandon the location. Bison can also be temperamental. They may be relatively docile and easily handled one day and dangerously intractable the next. The first sighting of a person afoot or on horseback may go almost unnoticed one day; on a different day, it can stampede a herd that may not even pause and look back for a kilometer or more. Such seemingly erratic behaviors are probably responses to conditions beyond the perceptions of the frustrated humans affected.

Arroyo Bison Traps

One feature characteristic of a large part of the plains is dry arroyos, within which water runoff eventually leads to the formation of perpendicular knick points, or headcuts. In unconsolidated sediments, these headcuts continually migrate upstream as new ones form downstream; in many dry arroyos, one can look upstream and see a sequence of them. Some are high enough to form barriers to animals moving upstream; and if the stream banks on both sides approaching the headcut are too steep to allow them to exit, this combination of features forms a natural trap. The only way that animals in such an arroyo can escape is to turn around and leave as they had entered. If the animals can be driven into the arroyo bottom where the banks are lower and not as steep, the hunters then need only to keep the animals moving up the stream channel until steep banks and a perpendicular headcut prevent their escape. This kind of landform provided an effective means for trapping bison in prehistoric times. Various configurations of them used for that purpose are found throughout the plains. Perhaps their greatest attraction was their ubiquity: it was almost always possible to locate one in the vicinity of a bison herd, and it required little or no modification—the steep headcuts and arroyo banks eliminated much of the need for corral posts and log superstructures to contain the animals.

In some of these arroyo traps, two closely related but distinct processes resulted in all or large portions of the skeletal remains being preserved in their original positions. If the kill occurred during a period of geologic aggradation, the carcasses sometimes collected alluvial and colluvial sediments as the bones were covered to varying depths. In rare cases, the geologic regime subsequently changed to one of degradation and the arroyo channel took a new course, leaving the bone bed intact. This was the situation at the Colby Mammoth Kill, described in the previous chapter. At other times, when kills were made during periods of geologic downcutting, the bones could collect enough alluvial material as the water ran off to form obstacles large enough to force the stream to move laterally, also leaving the bone beds intact. If later downcutting continued, part or even all of the bone bed might be preserved and its edge might become visible in the bank above the present arroyo bottom. To interpret

FIGURE 13. The Hawken site in northeast Wyoming: *a*, bottom of the old arroyo; *b*, upper bison bone bed: *c*, location of the headcut in the old arroyo not yet exposed. (From Frison, Wilson, & Wilson 1976: 29.)

how such animal kill sites were formed, one must accurately reconstruct the landforms involved—a requirement that provides a strong justification for expertise in what has come to be recognized as the subdiscipline of geoarchaeology. Lacking accurate identifications of past landforms, the archaeologist attempting to describe a past procurement strategy will often err.

An outstanding example of a bison trap in a dry arroyo is located in northeastern Wyoming toward the western edge of the Black Hills. Deep arroyos originating on steep slopes drain into a large open basin several kilometers across; its lush grass and ponds offer ideal conditions for attracting large mammals such as bison to a place where they can be consolidated into herds of the proper size for driving into traps. As gradients of the arroyos decrease toward the interior of the basin, bottoms widen and flatten, and concentrated runoff water produces lush grass. Cattle at present and bison in the past were attracted to these locations; in at least two arroyos, the latter were driven upstream until a headcut and nearly perpendicular banks formed barriers to further travel. In the process described above, bones left at a kill location collected alluvial and colluvial material, forcing lateral cutting of the arroyo. The later downcutting of a new channel left partial remains of nearly a hundred bison well preserved in a segment of the old channel. Known as the Hawken site (figure 13), this trap contained three discrete levels of bison bones; these partial remains of at least eighty animals, killed in late fall or early winter, have been dated to about 6,600 years ago. Metric analysis of the skeletal remains identified them as *B. occidentalis*, intermediate in size between the early Holocene and present-day bison. Large numbers of broken and complete projectile points, stone tools, and human-modified bones provide clear evidence of a human kill (Frison, Wilson, and Wilson 1976).

A similar arroyo bison kill occurred in the Powder River Basin in northern Wyoming; dated at about 2,600 years ago, it involved *B. bison bison* (Frison 1968b). Enough of the landform present at the time of the kill remains to allow its accurate reconstruction at the time it was used as a trap (Mann 1968). In this case the headcut in the arroyo, known as the Powder River site, was relatively durable, because it was capped by a layer of sandstone that resisted erosion. However, only a small remnant of a much larger bone bed (figure 14) managed to survive geologic activity that occurred after the animals were killed.

The Carter/Kerr-McGee site, an arroyo trap used for about 2,500 years during Folsom and into Cody Paleoindian times, is located in the Powder River Basin in eastern Wyoming (Frison 1984). The area contains coal beds up to 30 meters thick and numerous large strip mines. Many coal veins burned in the past, probably during the late Pliocene or early Pleistocene; the resulting land subsidence has left large depressions, many containing shallow lakes. One of these depressions is about a kilometer in diameter and has a maximum depth of about 20 meters. Runoff water draining into the depression formed arroyos around its perimeter, and one of these is about 12 meters deep with a flat bottom and steep sides.

FIGURE 14. Bison bone bed at the Powder River site in northern Wyoming. (From Frison 1991b: 196.)

Water comes to the surface in the arroyo bottom during spring and much of the summer to produce lush grass that attracts animals. As the arroyo gradient increases, headcuts formed animal traps in the past, and erosion has continued to form new headcuts to this day. Repeated cycles of this process in the arroyo at the Carter/Kerr-McGee site since its last known use in Cody Paleoindian times eliminated all but a small remnant of what was apparently a large bison trap; given the present rates of erosion, it is likely that only the site's discovery in 1974 has preserved any part of it.

The Agate Basin site in extreme eastern Wyoming is characterized by arroyos continually encroaching into an old, slightly elevated land surface. In situ bison bone beds date from Folsom times, beginning about 10,800 years ago; through Agate Basin times, around 10,300 years ago; and up to Hell Gap times, about 10,000 years ago (Frison and Stanford

FIGURE 15. Modern analogue of an arroyo bison trap. For scale, note the person standing in the bottom. (From Frison 1982a: 268.)

1982a). At present, the main arroyo presents an ideal route for driving animals attracted by lush grasses in the arroyo bottom up against head-cuts, as the gradient increases sharply—precisely as the arroyo traps present in Paleoindian times are believed to have been laid out. However, headward erosion since then has advanced further into the elevated land surface and removed the original headcuts that formed the traps used by the Paleoindian hunters. Although the site area is currently devoid of trees, the presence of certain small mammal remains indicates some were present during Folsom times (Walker 1982: 303) and thus could have been used to improve an arroyo trap. At present, at least two arroyos with headcuts and steep sides in the immediate Agate Basin site area would be excellent traps without any modification (figure 15).

An arroyo was used in an entirely different way to trap bison at the 10,000-year-old Olsen-Chubbuck bison kill on the plains of eastern Colorado. In this instance, a herd of *B. occidentalis* was stampeded into a deep, narrow, steep-sided arroyo, killing an estimated 190 animals. Complete bison skeletons in the arroyo bottom, killed by the mass of animals above and buried too deep to allow access to the hunters, leave no doubt as to the direction of the stampede, which was at right angles to the arroyo (Wheat 1972). It was first described as a spring kill, but later examination of tooth eruption of the calves established their time of death as late summer or possibly early fall (see Frison 1991b: 279–81).

Three separate kill episodes are well established at the Cooper site, an arroyo kill of Folsom age in northwest Oklahoma (Bement 1999). Lateral cutting of the Beaver River stopped just short of eliminating the site, sparing part of it that contained well-preserved skeletal material. A large male skull with a red ochre, lightning-like symbol painted on the frontal bone that was placed on top of the lowest of the three bone beds provides a strong argument for ritual activity connected with the site. In 2002 a Clovis-age bison kill was confirmed a short distance away in the same arroyo system, at the Jake Bluff site 34HP60. Future investigations there will undoubtedly expand our knowledge and understanding of arroyo bison kills on the plains into the Clovis period. A date of 10,750 B.P. on a bison tooth, if reliable, could indicate the transition between Clovis and Folsom bison procurement for that area (Bement, Carter, and Buehler 2002). Besides this site, the only confirmed Clovis bison kill is the Murray Springs site in southeast Arizona (Haynes 1993).

Preservation of bison bones in the arroyos described above is the result of accidental geological processes, and they most likely represent only a very small percentage of the original total number of kill sites. One must also keep in mind that those geological processes cause ceaseless alterations. Over a period of years, or even after heavy runoff from a single storm or spring snow melt, a trap may be rendered unusable. Such shifts in dry arroyo systems seldom leave the archaeologist much good data. However, from these sites we can derive a picture of one widespread bison procurement strategy used during several thousand years of plains prehistory.

Some arroyo bison traps were of a variant form that is less visible archaeologically, probably because their location left them more vulnerable to loss through geologic processes. They relied more on shallow arroyos and on longer and steeper slopes. There are three of these presently known to me, one in southeast Montana and two in northeast Wyoming. The kill locations are close to the shallow heads of arroyos, at the termination of long, steep uphill climbs for the animals. Their striking similarity suggests that the locations had to be carefully chosen to be successful. One of these, the Cache Hill site (Miller and Burgett 2000; see figure 16), is about 300 years old, falling close to the end of the Late Prehistoric period. The other two sites have only been subjected to preliminary investigation, but the projectile points recovered in them are the same as or similar to those in the Cache Hill site. The bison in the latter

FIGURE 16. The Cache Hill bison kill, northern Wyoming. (Photo by Mark Miller.)

site were killed in late fall or early winter, and there is evidence of multiple kills—either clustered during a single year or at about the same time during different years. The Oshoto, Wyoming, site about 35 kilometers northeast of Cache Hill and the Marshall Lambert site near Ekalaka, Montana, about 190 kilometers away in the same direction, both unpublished, have more than one component and projectile points indicative of the Late Prehistoric period. Multiple use demonstrates that this procurement strategy was successful.

One advantage of this type of site over a jump was that a large herd of animals was not required; nor did hunters need to stampede them to their death. A cow and calf or two dozen or more bison could have been collected and then driven up from the bottom of an arroyo for a kilometer or more until they were close to the top of a ridge at the arroyo's head. The animals would have been in good condition in the fall; by the time they were just short of the ridgetop, they would have been winded and less aware of danger. This would have been the optimum moment for hunters waiting out of sight behind the ridge to confront the animals and kill at least some of the herd. This explanation seems to take account of the location of the bone beds in these sites in a way most congruent with bison behavior. Weather might also have played a role in this type of procurement. In fact, a heavy early fall snow might have worked to the advantage of the hunters.

Such sites may have been used earlier, but if so, erosion has destroyed the evidence. At the Upper Twin Mountain site near the town of Kremm-ling, Colorado, located in an intermontane basin known as Middle Park at an elevation of 2,548 meters, partial remains of at least fifteen extinct bison were recovered on a slope in what ultimately proved to be the trough of an old geologic slump scar about 75 meters long and 1.5 meters deep (Kornfeld, Frison, et al. 1999). Radiocarbon dating put the bones at 10,200 to 10,400 years old, and associated stone tools are technologi-cally and morphologically similar to the Plainview and Goshen types (see Frison 1996; Irwin-Williams et al. 1973). Judging from tooth eruption and wear, the animals were killed in the late fall or early winter.

The slump is not of a size or shape which would have formed a bison trap, and there is no evidence of postholes for a corral; it is difficult to envision driving bison here. However, over the top of the divide, a short distance from the site, lies a steep trail leading from a flat area at the bot-tom to the ridgetop directly above and near the slump scar. A plausible strategy for killing the animals at the Upper Twin Mountain site would have been to have one group of hunters pursue them to the top of the di-vide where waiting hunters could ambush them. Gravity, runoff water, carnivores, and scavengers could subsequently have moved butchered car-cass parts downslope into the slump, where they were trapped and ulti-mately buried. The paucity of projectile points and tools in the bone bed is another indication that it was probably not the actual kill location. The site is very difficult to interpret with any sense of surety.

Yet because these bison died in the late fall or early winter, we can say with confidence that late Pleistocene or early Holocene bison were able to survive the winter at high altitudes. Before this discovery, I had as-sumed that bison in Middle Park would have spent the winter months in lower elevations further north, probably along the North Platte River.

The bone bed at the Goshen-age Mill Iron site in eastern Montana may be the oldest one known on the Northern Plains. Parts of at least thirty-five animals (probably *Bison antiquus*) with radiocarbon dates of 11,000 years ago were recovered in the remnant of a bone bed about 2 meters below the top of a small butte. Intense erosion destroyed too much of the surrounding area to allow geologists to identify old landforms and reveal a plausible procurement strategy (Frison 1996).

Sand Dune Bison Traps

Both active and inactive areas of sand dunes are common on the plains. Vegetated sand dune areas are relatively stable; but if something destroys

FIGURE 17. Bone bed at the 10,000-year-old Casper bison kill site.
(From Frison 1974: 63.)

the root systems in even a very small area, the wind will begin to move
the sand. The results can be long, narrow, and deep parabolic sand dunes
with steep sides that can be utilized as animal traps. Such was the case
at the Casper, Wyoming, Paleoindian bison kill site of Hell Gap age, or
about 10,000 years old (see Frison 1974). A major problem in sand dune
sites is that archaeological materials are repeatedly exposed, reburied,
and reexposed by wind. Bone once laid bare to the elements rapidly de-
composes, and lithic materials no longer remain in their original context. The bison bones here were covered to a depth of nearly 9 meters a
short time after the animals were killed; they were not exposed again until earthmoving equipment removed the overburden on top of the bone
bed. The remains of approximately 100 *B. antiquus* killed and butchered
by Hell Gap–age hunters lay in their original position in what proved to
be the bottom, or trough, of an old parabolic sand dune (figure 17).

The trough was formed on a cobble surface impervious to water; after it was covered with sand, a pond formed over the bones. Carbonates
precipitated from the pond water preserved the bones unusually well.
The circumstances in which the animals were killed, their rapid burial
without subsequent exposure, and the concentrated carbonates that enhanced preservation resulted in a 10,000-year-old *B. antiquus* bone bed.
That the animals were killed in late fall or early winter was probably an-

FIGURE 18. Parabolic sand dune in the dune field near the Casper site.
(Photo by author.)

other aid to preservation. The discovery of a parabolic sand dune used
as a bison trap revealed a hitherto unknown and unsuspected strategy
for procuring late Pleistocene bison. Taphonomic study of the Casper site
bison added significantly to our knowledge of the anatomy of extinct bison. At least two parabolic dunes within a short distance of the Casper
site at the time it was investigated would be ideal for trapping bison (figure
18). However, it has been more than a quarter century since the photo-
graph reproduced in figure 18 was taken, and the configuration may have
changed by now.

The Cattleguard site in the San Luis Valley of south-central Colorado,
where bison were killed, butchered, and processed, dates to the Folsom
period, between 10,000 and 11,000 years ago, and is situated in an area
of extensive sand dunes (Jodry and Stanford 1992). There is no reason
to suppose that a parabolic dune was used as a trap. Here the Folsom
deposits were near the surface, and the history of their formation is not
as clear as at the Casper site. The Finley Paleoindian site in western
Wyoming is similar (Moss et al. 1951). The bison bone bed there yielded
partial remains of fifty-nine animals; however, because the site had been
looted earlier by artifact hunters and the bones they had left exposed deteriorated rapidly, the actual number of bison originally present in the
bone bed will never be known The bones were in shallow sand deposits,
and there was no convincing evidence that the location was ever the
trough of a parabolic dune. Sand dune fields cover large areas of
Wyoming (see Kolm 1974), and prehistoric bison hunters undoubtedly
used other of their features as aids in hunting. As is true of arroyo traps,

expertise in geology is needed to identify and reconstruct the landforms utilized and the processes involved in their formation. Sandy areas attracted prehistoric people, and not all of the attraction was connected with procuring animals. A sandy floor is much preferred to one of clay, especially in wet weather.

Elongated parabolic sand dunes analogous to the one used at the Casper site are prominent features throughout the area of vegetated sand dunes surrounding the site. The sides of the dunes are at sand's angle of repose, about 33 degrees, and the top meter or so forms a perpendicular barrier held in place by the extensive root system of native grasses that would make escape for the animals over the top of either side of the dune impossible. The nose, or leeward, ends of these dunes are usually less steep than the sides, but the sand—especially for large, split-hoofed animals such as domestic cattle or bison—impedes travel, giving human hunters a definite advantage. Parabolic dunes can be efficient as animal traps, though strong winds that move large amounts of sand can rapidly change their configuration and make them useless.

Bison Jumps

Driving animals over high perpendicular cliffs so that large numbers might fall to their death and forcing them down steep slopes or over low cliffs into constructed restraint pens where they could easily be killed were long-established strategies for bison procurement on the plains. Folsom hunters apparently drove bison over a perpendicular bluff more than 10,000 years ago at the Bonfire Shelter site along the Pecos River in Texas (Dibble and Lorrain 1968). It would be almost 5,000 years before the next convincing evidence for bison jumping appears—when intensive bison jumping began and continued well into the historic period at the Head-Smashed-In site near Lethbridge, Alberta (Reeves 1978b). The long gap in the historical record may reflect a sampling problem, and perhaps future discoveries will provide evidence that the practice continued during that time.

Bison jumps are the most visible type of procurement site because of the high cliffs, the drive lines leading to the edge of the cliffs, and the bone deposits along the base of the cliffs. There are numerous variations; some perpendicular stone cliffs last indefinitely but those of unconsolidated sediments deteriorate rapidly. Animals were sometimes driven over bluffs into temporarily dry river- and streambeds that later flooded, removing the bones (Malouf 1962b: 54–56). Corrals or similar impediments were placed at the base of steep dirt banks or precipices that were not so high that the impact from the fall would kill enough animals outright. In those cases, restraining structures were needed to prevent the escape of both crippled and uninjured animals.

Although their principles of operation are similar, every bison jump is unique. The Head-Smashed-In site exhibits classic bison jump features, including extensive drive lines extending several kilometers from animal gathering areas and a jump-off with a perpendicular drop of sufficient height to kill many animals immediately and cripple many others (so, too, at the Kobold Buffalo Jump; figure 19). The Logan Buffalo Jump on the Madison River near Three Forks, Montana, has a 10-meter vertical drop below the jump-off, followed by another 30 meters of steep slope, features that together make it extremely unlikely that any animals driven over the edge survived (Malouf 1962a). This site was used many times over a period of several thousand years, as confirmed by deep, stratified deposits of bison bone with projectile points specific to several time periods. The Glenrock Buffalo Jump in central Wyoming is reminiscent of the Logan Jump except for having a shorter time depth. Near the town of Cayley, Alberta, a vertical cliff now 6 meters high and possibly twice that height at the time of its first use is known as the Old Women's Buffalo Jump. Deposits below the jump-off are at least 6 meters thick in places; along with the accompanying artifacts, they indicate different episodes of use over the past 2,000 years (Forbis 1962a, 1962b). Fortunately, this site was never mined for bones for use as fertilizer, as were most large bone beds associated with jump sites (considerable quantities of bone were removed in the twentieth century; see Davis 1978).

Most investigators prefer to restrict the designation *buffalo jump* to those sites with cliffs high enough to kill or seriously wound a large share of the animals stampeded over the edge (see Malouf and Conner 1962). The number of known buffalo jumps is well into the hundreds; some reflect intensive, long-term use, while others may have served this purpose only once. The center of this activity was Montana; many jumps can also be found in Alberta and Wyoming, with fewer in Saskatchewan, Manitoba, the Dakotas, Idaho, and Colorado. The obvious requirements were topographic, including the proper approaches, along with the presence of adequate numbers of animals in locations favorable for driving them to the jump-off. Yet another consideration was access to the dead animals. In countless places animals could have been stam-

FIGURE 19. The Kobold Buffalo Jump, south-central Montana. The arrow marks the drive lines leading to the jump-off. (From Frison 1978: 209.)

peded over high cliffs but would have been impossible to retrieve from below.

H. P. Lewis of Conrad, Montana, which is located in the heartland of Northern Plains buffalo jumping territory, was an early-twentieth-century artifact hunter attracted to these sites because of the wealth of projectile points found in the bone deposits. Along with his collecting activities, he developed a strong interest in seeking out the stone piles that formed the drive lines and tracing the paths the bison took as they were driven to the jump-offs. He wrote of his experiences in locating buffalo jumps and in digging for artifacts before the development of modern archaeological investigations; though he never completed his narrative, it should be read by anyone seeking information on buffalo jumps (H. Lewis 1947).

The reader of Lewis's account soon realizes that his interests went beyond collecting artifacts. It appears that he began to locate jump sites, as marked by bison bone deposits, and then started to trace back the drive lines to their beginnings. He was particularly impressed with one jump site—the Two Medicine kill, located in northwest Montana just east of Glacier National Park; he spent some time locating the several kilometers of drive lines leading to the jump and postulating the route over which the animals were driven. He describes a hollow, close to the jump-off, with low ridges on both sides through which the animals were driven and continues:

> There is not the slightest sign to a stampeding herd to indicate anything to obstruct their run until they are fully upon it. In fact, a hundred animals could cross the low ridge and be out of sight of the remainder of the herd. No more diabolical arrangement for wholesale slaughter could be conceived than this deceptive spot. Small wonder the unsuspecting, and greatly excited buffalos could be driven over the point by hundreds. (H. Lewis 1947: site #19-3)

Lewis's narrative conveys his deep understanding of animal behavior. It is unfortunate he was not able to pass on more of his knowledge to present-day archaeologists. Though there is not space here for even a brief discussion of other Montana buffalo jumps, a few should be mentioned: Ulm Pishkun (Davis 1978) and Wahpa Chu'gn (Davis and Stallcop 1976) in northern Montana and Antonsen (Davis and Zeier 1978) in the southern part of the state.

Chugwater is a small town in eastern Wyoming, located along a stream in a valley with steep bluffs on both sides. Local folklore claims the name was derived from the sound, or "chug," that bison made when they hit the ground at the bottom of the bluffs. According to one account, on one occasion several Native Americans on horseback stampeded a herd of several hundred bison over one of the bluffs. If this actually happened, it would have occurred around the middle of the nineteenth century. Many wagonloads of bison bone were also said to have been later collected at this location and sold for fertilizer. Several bones remaining close to the base of the bluff suggest that this may have been the case (Trenholm 1970: 66). As a location for a bison jump, it appears to be ideal: until almost at the jump-off, the land surface blends so well with the opposite side of the valley that the two appear continuous. Although the topography is different from that of the Two Medicine jump, it is equally easy to envision a herd of stampeding bison unaware of the jump-off until they could not avoid it. I could find no evidence of bison jumping orchestrated by hunters on foot, though that absence may not be very significant: the surface at the bottom of the bluff lies on the flood plain of Chugwater Creek and is often scoured out by floods.

An unusual bison jump in extreme southeast Idaho utilized a large, flat-topped butte with several narrow extensions that, once the animals were stampeded out onto them, allowed them little choice but to go over the edge. Its configuration resulted in the name "Five Fingers Bison Jump" (Agenbroad 1978a).

Variants of Buffalo Jumps

Covering an area of approximately 20,000 square kilometers, the Black Hills of South Dakota and Wyoming rise up in stark contrast to the surrounding plains. Steep slopes, deep arroyos, and increased vegetation now provide ideal bison habitat and, judging from the archaeological record, did so well into the past. Such locations are irresistible to bison. In the northeast corner of Wyoming is an area of karsts, or solution caverns, in thick gypsum beds; one of these karsts provided a unique variant of a bison jump, known as the Vore Buffalo Jump (figure 20). Situated in relatively open country of moderate topography, it consists of a nearly round, steep-sided depression about 30 meters across at the top, 15 meters deep, and 25 meters in diameter at the bottom. At least twenty-two separate, stratified levels of bison bone are present from about 1.2 to 3.7 meters deep in the bottom of the karst, laid down between 1500 and 1800 C.E. An unusual aspect of the Vore site is that unlike with other jump sites, its form allowed animals to be driven into it from any direction. Though most of the drive lines were lost to farming and road construction, the segments that remain suggest that animals indeed approached it from several directions.

The Vore site shares with the Two Medicine and Chugwater sites another feature that enhances its value: it is difficult to see the jump-off until one is nearly at the edge. Projectile points in large numbers throughout the bones strongly indicate that the trip down the steep slope to the bottom of the karst was rarely fatal and most of the animals had to be dispatched by the hunters using bows and arrows. As the most recent bones are from about 1800, it is possible the last use of the Vore Buffalo Jump was by hunters using horses. Judging from the tooth eruption of calves, one kill episode occurred in late spring. Less than 10 percent of the site has been excavated, but it is estimated that between 10,000 and 20,000 animals were killed there (Reher and Frison 1980; see figure 21). The site is well protected and skeletal material is well preserved, promising rich yields for future researchers.

The Willow Springs Buffalo Jump near Laramie, Wyoming, is a classic example of a location at which bison were driven over a low, perpendicular bluff (2.5 meters high) and into a corral. Postholes and bases of several posts, along with fragments of pitch pine logs, indicate a sub-

FIGURE 20. The Vore Buffalo Jump in northeast Wyoming. (From Reher &
Frison 1980: 8.)

stantially built, rectangular enclosure measuring approximately 15 meters by 21 meters, with rounded and strongly braced corners and its long
axis parallel to the drop-off. The structure contained burned bison bone
up to 60 centimeters in depth. Burned bone is found in other kill sites,
and no completely satisfactory explanation has yet been found: the fires
could have been intended to eliminate the debris resulting from earlier
kills and the tall vegetation that thrived on the nutrients from the body
fluids of animals killed the previous year. Thousands of bow and arrow
projectile points were recovered among the upper bone levels, and dart
points in the bottom level indicate intermittent use for about 2,000 years.
As at the Vore site, the sheer number of projectile points suggests that
the jump-off failed to kill many of the animals (Bupp 1981; Frison
1991b: 231).

Part of the Crow Indian territory during Late Prehistoric and Early
Historic times included the Big Horn Mountains in northern Wyoming.
The eastern slopes of the Big Horns were prime bison country, with clear
streams and lush grass. Two known bison kill sites believed to be of Crow
origin utilized drive line systems: animals were forced over steep slopes

FIGURE 21. Bison bone bed at the Vore Buffalo
Jump. (From Frison 1978: 241.)

into restraining structures at the bottom. The bone bed of one of these,
the Big Goose Creek site (Frison, Wilson, and Walker 1978), was nearly
removed by a major flood; the stratified levels of the small part left intact indicate repeated use. Of greatest interest is a system of drive lines,
continuous except for short gaps caused by irrigation ditches and roads,
that extend for nearly 2 kilometers into a large gathering area on the
mountain slopes.

    The drive line markers are stone piles, each containing from fifteen to
twenty boulders and cobbles, placed about 3.5 to 7.5 meters apart and
forming nearly continuous lines. In my opinion, these markers were meant
to indicate the limits within which the hunters needed to keep a herd of
bison while maneuvering them to their final destination, a steep bank of
about 55 degrees that was about 14 meters high. The stone piles may
also have served the same purpose, but in another way. Two researchers
studying the drive lines at Head-Smashed-In raised the possibility that
additions to the cairns could have increased their visibility and "directly
affected the behavior of the bison," arguing that the cairn lines at times
"must have had to function on their own in directing herds being man-

aged for jumping" (Brink and Rollans 1990: 160). Historical accounts suggest that the additions to cairns could have been perishable materials such as brush or bison chips for which there is no remaining evidence. One of the more authentic eyewitness reports on drive lines is that of Peter Fidler of Hudson's Bay Company, who traveled through southern Alberta in 1792–93. He described a group of Peigan who were still driving bison although they were using horses (quoted in Forbis 1962a: 63; see Fidler n.d.). Besides being functional, the cairns and cairn lines may also have had ritual significance.

Upstream from the kill area at Big Goose Creek about 200 meters and on the same side of the stream is a large campsite that was at first assumed to be directly associated with the kill. Only one of the kill area components produced enough intact mandibles from immature animals to indicate the time of year of operation: midfall, in late October and early November. The camp area produced a large quantity of bison bones but only two mandibles, both from mature animals whose age is difficult to determine accurately and both insufficient to hazard a guess at what the time of year they were killed. However, the same area did yield maxillae of thirteen animals with intact tooth rows, several from immature animals (apparently available because of how the skulls were processed by the hunters). These indicated that the animals did not die at the same time of year as those found at the kill site; instead, they were individuals killed throughout the winter months.

Another line of evidence from the camp area supports this interpretation. Bones from twelve different fetuses form an unbroken growth series estimated to have lasted from November through February; they indicate gradual killing in the camp area quite distinct from the catastrophic kill at the jump area. This information complicates the interpretations of the Big Goose Creek site. The processing of the animals from the kill site must have occurred elsewhere—highly possible, given the distance between the two and the changes in course of the stream over time. If we assume that the two areas are directly connected, then the most likely explanation is that the hunting group occupied the camp during the winter and killed an occasional bison while drawing on the stored meat surpluses acquired during the fall kill. It is also possible that the kill and camp are not directly connected and represent different cultural groups and different years of use.

Another unusual feature of the Big Goose Creek site is a narrow, 15-meter-deep, steep-sided arroyo immediately adjacent and at right angles to the jump-off. At least five and possibly more complete bison skeletons were tightly wedged into the bottom of this arroyo, and none showed any evidence of butchering. These animals may well have been crowded into the arroyo during the final stampede, and left unrecovered because of their location's inaccessibility. This is one of a very few prehistoric sites I am aware of with unequivocal evidence of failure to utilize dead animals; it is strongly reminiscent of the findings at the bottom of the arroyo at the Olsen-Chubbuck site (Wheat 1972), where the pile of animals at the top obstructed access to the ones at the bottom.

At the Piney Creek site (Frison 1967b), nearly all of the bone bed remained intact at the bottom of a 48 degree slope 13.5 meters high. Although any restraining structure is now rotted away, its former presence is strongly indicated by the abrupt edges of the bone bed and the large numbers of projectile points among the bones. However, unlike the Big Goose Creek site, Piney Creek appears to have been a onetime operation. Nearly all of the drive lines were eliminated by construction except those close to the edge of the jump-off. A large processing area is immediately adjacent to the kill.

Perhaps the best-known example of this kind of bison procurement complex is the Boarding School Bison Drive site near Browning, Montana (Kehoe 1967). As at the Piney Creek and Big Goose Creek sites, bison were driven over a steep incline into a corral at the base. An estimated 250 bison were taken in two separate drives that occurred after 1600 c.e. A Blackfoot Indian in 1947 told a researcher that he had witnessed hunters using horses to corral bison (Ewers 1949: 359). His details of corral construction fit well with the Boarding School site and the postulated structures at the Piney Creek and Big Goose Creek sites. Cottonwood posts that extended about 2 meters above ground were set in postholes at the Boarding School site. Cottonwood logs were then tied to the posts with rawhide to form the enclosure. Poles were also laid side by side and parallel to the slope leading to the enclosure. These were then covered with manure and water, which froze and ensured that the animals lacked adequate footing to climb back up the steep incline to escape.

A similar procurement strategy was present more than 4,000 years earlier, soon after the reappearance of bison following the Altithermal period. A well-preserved segment of a bison kill, the Scoggin site in central Wyoming (Lobdell 1973; Frison 1991b: 193–94), consisted of a thick bone bed between postholes. Colluvium from the slope covered the site to a depth of 0.75 meters. The postholes were connected by a low wall of dry-laid flat stones at the base of a steep talus slope about 7 meters high, formed by a thin layer of cap rock overlying softer clay deposits; together they formed a substantial fence (though not an enclosed structure), whose ends extend upslope at right angles for a short distance. One corner of the fence was still intact, and extra postholes indicate added reinforcement at a critical spot. Bison long bones had been forced into the downslope sides of several postholes, presumably to straighten posts leaning because of the intense pressure against them exerted by the animals.

The momentum of animals stampeded over the edge would have carried them down the slope until they encountered the fence. The only possibility of escape was to go back the same way; but the hunters would have had a definite advantage over the disoriented animals milling around below them. Large numbers of dart points indicate how the animals were killed. The site is strategically located at one end of a natural and well-traveled corridor through a high, narrow ridge several kilometers long that presents a substantial barrier to animals wanting to cross it. Outside the fence was a stone boiling pit that contained two large male skulls which were well within the size range of modern bison. A radiocarbon date of 4550 B.P. ± 110 years (RL-174) is strong evidence for the appearance of *B. bison bison* approximately 2,000 years after the intermediate bison found at the Hawken site (Frison, Wilson, and Wilson 1976).

Bison herds could be driven or stampeded down into the Piney Creek, Big Goose Creek, Boarding School, and Scoggin sites today; but without substantial restraining structures at the bottom to hold them, hunters could expect little more than a dead animal or two and a larger number with broken legs—small payback for the effort expended in building the complex, gathering a herd, and maneuvering the animals into proper position. The ones that escaped would need at least a day or so to recover before there would be any chance of trying again. Depending on the number of animals available in the area, the best strategy probably would be to find another herd. However, the remnants of structures at the sites combine to indicate a much more successful bison procurement strategy, used by pedestrian hunters beginning at least 4,500 years ago and continuing until just before the introduction of the horse in early historic times.

Bison Corrals

Corrals or pounds required artificial structures with many features similar to modern cattle corrals. One of the most accomplished prehistoric pedestrian bison hunting groups on the plains was present in the Wyoming area between about 2,000 and 1,700 years ago; it takes its name, Besant, from a site in the Besant Valley in Saskatchewan, Canada (Wettlaufer 1955). Similarities in weaponry suggest that its members may have acquired bison hunting techniques there and then moved south and west into the plains. Besant sites are widespread in Saskatchewan, Alberta, Montana, and Wyoming. Two in Wyoming that are unusually well preserved reveal high levels of expertise in bison procurement; these skills can probably be generalized to all large Besant bison procurement locations.

One of these sites is in an isolated area in the Powder River Basin on the plains of eastern Wyoming, where an accident of geological activity resulted in the preservation of parts of a large bison corral and a structure with features believed to reflect associated ritual observances. Located in a meander bend of a dry arroyo that was aggrading at the time, the site was covered with up to 3 meters of alluvial deposits after it was abandoned. When the geologic regime reversed again, downcutting scoured out the old arroyo but left some of the meander bends intact, including the one that contained the Ruby Bison Pound (Frison 1971a; Albanese 1971). In the process, part of the bone bed was exposed, attracting the attention of local ranchers. Though the site is difficult to access, word spread rapidly that quantities of large projectile points could be found in the bones, and it became known locally as the "arrowhead mine." By the time I was able to visit the site, most of the bone bed had been removed. It was rumored that one "miner" undercut the bone bed so far that the bank collapsed; luckily, he was not alone, and his companions managed to rescue him.

Once the disturbed bone bed deposits were removed, postholes emerged and gradually began to form a pattern suggesting a former corral structure (map 4). The bottoms of several holes contained remnants of large juniper posts burned at the bottoms, indicating at least one method of separating them from their stumps. The other species present in quantity and probably also used in construction of the corral is cottonwood, which rots away within a year or so when left underground. Large numbers of trees of both species, living and dead, are plentiful in the arroyo at the site today.

As the corral area was peeled off in thin layers, earlier building stages were revealed, indicating more than a single year's use. The posthole pattern suggests paired posts with logs stacked between them, one on top of the other—a method of constructing livestock corrals commonly used today. That the structure was apparently placed deliberately on a slope

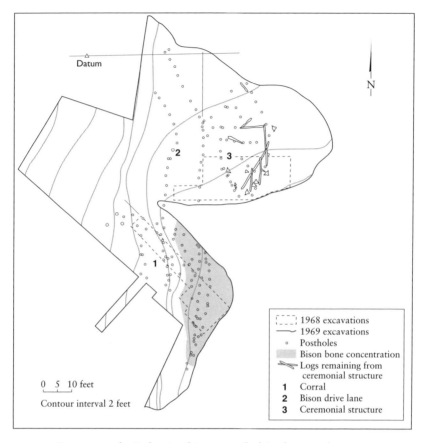

Datum

N

- - - 1968 excavations
—— 1969 excavations
∘ Postholes
▨ Bison bone concentration
🪵 Logs remaining from
   ceremonial structure
1 Corral
2 Bison drive lane
3 Ceremonial structure

0 5 10 feet

Contour interval 2 feet

MAP 4. Features at the Ruby site: bison corral, drive lane, and structure. From Frison 1971a: 78.

would puzzle today's cattle ranchers, but the prehistoric hunters were not interested in operations such as branding, dehorning, cutting, and castrating, which need to be carried out on level surfaces. If a cattle operation needed only to kill animals, the Ruby site corral would function quite well. The downhill side of a corral built on a slope is ideal for holding animals to be killed, provided the fence is strong enough. And because the upslope sides of the corral do not need to be as sturdy, considerable effort can be saved. In addition, this kind of structure allows the hunters on the elevated side of the enclosure a less obstructed view of the animals and thus better targets. Indiscriminate shooting into a crowded mass of animals, or bunch shooting, is a poor strategy regardless of the weaponry used, because it usually results in wasted shots and wounded animals. The experienced hunter knows to select one animal; only when certain that a lethal wound has been delivered to that animal does he turn his attention to another.

A narrow but deep arroyo removed part of the corral, but postholes appeared to continue beyond the bone bed; at that point, investigation required heavier equipment to remove the nearly 3 meters of alluvial deposits. Their pattern changed from paired posts to two separate lines of postholes about 3 meters apart, forming a drive lane that led to the corral entrance; about 10 meters outside the entrance was a bend. Postholes for the drive lane ended where alluvial deposits covering the site area ended; all further potential evidence was lost to erosion. But it is almost certain that the widening wings required strong fences to be extended for some distance beyond the drive lane entrance.

Numerous projectile points on the floor of the drive lane were found at its bend, indicating that the hunters on both sides were prodding the animals to keep them moving ahead. The hunters may also have intended to inflict lethal wounds while the animals were in the drive lane, well aware that these wounds would take final effect when the animals were inside the corral. This stage of the process was critical. Many persons unfamiliar with bison have seen the bend simply as demonstrating the hunters' inability to build a straight drive lane. However, personal experience in handling both cattle and bison has taught me that the animals are far more apt to balk and turn around in straight drive lanes than in those with bends, which obscure the opening to the corral until the last possible moment. If one animal does manage to turn around, the rest will follow; heading back to open country, they will be impossible to stop. It is frustrating, to understate the feeling (and I again speak from experience), to come this close to success and then have to watch the herd disappear into the distance. Equally frustrating is the knowledge that the animals will travel some distance before stopping and, for a day or so at least, will be wary of a repeated attempt at corralling. This corral was ideally located, because it was obscured from the animals' view until they were well into the drive lane and close to the entrance. The procurement complex seems to have been used intensely and repeatedly, probably over a period of several years.

Postholes and logs adjacent to the corral and drive lane but with no apparent structural connection to them proved to be the remains of a structure football-shaped in outline, 5 meters wide and 13 meters long, formed by intersecting arcs of circles with 10.2-meter radii and oriented nearly true north and south (see map 4). Its all but perfect symmetry suggests that the builders set two stakes 15.5 meters apart and then marked off the two intersecting arcs. The badly rotted remains of a juniper log just under 6 meters long lies on top of another log almost 5 meters long. The former is believed to have been a ridgepole extending from the center of the structure almost to the south end, while the latter probably spanned the structure at the center from east to west. Two shorter logs on top of the longest may have been rafters from the ridgepole to the east side. The positions of these logs indicate that the structure collapsed to the east, with the long ridgepole on top of the cross log and the two rafters on top of the ridgepole—evidence of a roof over the southern half but not the northern half. Postholes outlining the structure were dug perpendicular to the ground, and three short logs may be the remnants of aboveground parts of posts. If so (and this is highly conjectural), the sides of the structure could have been as much as 1.4 meters high.

There was no evidence of daily activities or bison butchering and processing inside or outside the structure. Three holes of small diameter in the southern half each contained the second or third thoracic vertebra of a bison, with the long dorsal spine in the hole and the opposite end above ground. One hole just outside the structure to the west contained four thoracic vertebrae with their dorsal spines pointing downward. A complete set of articulated cervical vertebrae were in another hole, with the first vertebra at the bottom. Although somewhat disturbed, eight male bison skulls with mandibles removed and noses pointed outward were placed around the south end; a single male skull was just outside the structure on the northeast side.

Without known exception, ceremonial activity accompanied ethnographic accounts of communal bison hunting, and the structure alongside the Ruby site corral can only be interpreted in this same context. David Mandelbaum (1940: 190–91) claimed that the building of a Cree pound was supervised by a shaman who, when the trap was in operation, placed his tipi by the entrance, where he invoked spirit helpers to ensure success. Similar accounts treat other historic Plains tribes (see Grinnell 1961: 12–14; Gilmore 1924: 209; Chittenden and Richardson 1905: 1028–29). The structure at the Ruby site confirms that ritual activity was associated with communal bison procurement nearly 2,000 years ago.

Besant bison hunters operated another large communal hunting complex in the Shirley Basin area of central Wyoming. Known as the Muddy Creek site (Frison 1991b: 208–11; Hughes 1981), it is located near the headwaters of a small, meandering stream that flows through several square kilometers of excellent grazing and gathering area. The site is hidden from view, situated at the bottom of a slope just over the edge of a flat, mesalike area that is about 4 meters higher than the corral. The corral itself was on a slope, but not as steep as the one at the Ruby site. As was the case at the Ruby site, artifact hunters had visited the site and taken away hundreds of projectile points before any archaeologists arrived.

According to one group that had dug in the site, the bison bones were within a few centimeters of the surface in places and exposed in others. Clearing away the area revealed a round corral structure about 13 meters in diameter with single posts instead of the paired posts used at the Ruby site. The posts also were larger than at the Ruby site, probably because larger timber was available in stands of lodgepole and limber pine nearby. It is much easier to obtain and prepare posts of pine than of juniper and cottonwood. Bison long bones were upright in several holes, next to the posts and facing out, just as at the Scoggin site, and presumably for the same reason: to straighten posts pushed askew by the animals inside leaning on them. There was no evidence of paired posts, but these large posts may have required that a smaller post only be tied to them, not buried, to hold the horizontal logs. Another possibility is that the horizontal logs may have been tied to the posts in the fashion of the historic Blackfoot corral described above.

At first it was believed that bison herds were driven upstream and into the corral using a fenced drive lane, as was the case at the Ruby site. However, no evidence of the postholes for such a fence could be located; instead, an area of artificially placed cobbles was found that formed a pavement on the flat to the west of the corral. Further investigation revealed postholes on the steep slope between the pavement and the corral with no sign of any opening. The top of the corral was level with the flat to the west, and a ramp had been built that dropped the animals into the corral, eliminating the need for an entrance.

The selection of this location demonstrates that the hunters thoroughly understood bison behavior and carefully studied the topography to locate a combination of natural features that would serve their purpose. From a distance, the terrain seems almost flat, but this appearance is deceptive. The corral is in a depression, which hides it from the animals' view until the last possible moment; moreover, the boulder pavement was critical to the operation. At the ramp entrance, the lead animals would have attempted to stop by bracing all four feet; were it not for the pavement, they would have forced their hooves into the soft clay and disrupted the drive. A thin covering of soil was very likely spread over the

FIGURE 22. Bison bone bed at the Horner site, northwest Wyoming. (Photo by author.)

cobbles to make them less obvious and, as at the Blackfoot corral, water (or ice, if the weather were cold enough) would have made the clay slippery. Substantial fences leading to the ramp would seem to have been necessary, though no evidence of postholes was present; indeed, the hardpan just below the loose surface soil makes the digging of postholes almost impossible. A well-braced fence would have sufficed, however.

No evidence of a religious structure such as the one at the Ruby site could be located here, but we must keep in mind the locations' very different layouts. There, a structure immediately adjacent to the drive lane was obscured from the animals' view; here, a religious structure alongside the entrance of the ramp would have been in plain view of the animals being driven in that direction and thus unsuitably located. It is possible (though this suggestion is purely speculative) that as part of the labor-intensive construction of the ramp at the Muddy Creek corral, a space for the shaman was left beneath it and between the posts holding it up. There he could perform his rituals of calling in the animals.

Bison corrals probably date to the Paleoindian period. At the 10,000-

FIGURE 23. Nearly complete bison skeletons at the Horner site, which
appeared after the disarticulated bones shown in figure 22 were removed.
(From Frison & Todd 1987: 141.)

year-old bone bed at the Horner site (Frison and Todd 1987), the partial remains of more than sixty animals killed in late fall or early winter
were at first thought to be bones discarded after the flesh was stripped
from animals cut into units small enough to be transported from a kill
to a campsite for further processing (figure 22). However, after we removed the individual bones, several nearly complete carcasses appeared
(figure 23), and the bone inventory demonstrated that almost no bones
had been removed from the bone bed. It is highly unlikely that animals
this large were being killed in one place and moved as entire carcasses
to another location for butchering. We concluded that the animals had
been killed and butchered in this location, and the flesh removed to another location for processing. This interpretation was further supported
by an assemblage of broken and complete projectile points and tools sug-
gestive of killing and butchering, but a lack of tools and features in the
bone bed that would indicate meat processing.

The bone were well enough preserved to reveal that there was no intensive filleting of bones and very few were broken to open marrow cav-

ities, consistent with light butchering and relatively low utilization of the potential food resources. Perhaps the hides, which make surprisingly warm coverings, were as desirable as the meat products. Unlike a domestic cow hide, a bison hide with a heavy winter coat of hair is ideal for winter nights when temperatures dip into the double digits below zero. A strong possibility also remains that most if not all of the carcasses were stripped and left intact and that the disarticulation and scattering of bones may have been due to activity by carnivores and scavengers after the hunters left or by natural causes before they were covered and preserved by alluvial and colluvial deposits.

The edges of the bone bed terminated abruptly, suggesting that the animals were contained in some way. The bones were in a shallow drainage channel about 0.5 meters deep, far too shallow to contain bison. However, the height differential could benefit a fence. The compacted cobble level precluded the excavation of postholes to any depth, but a well-braced structure of heavy logs would have sufficed. The steep slopes of the nearby stream have heavy tree growth today, and trees were most likely readily available 10,000 years ago, when climatic conditions were similar. The drainage channel in which the bone bed lies continues for some distance to the southwest, and would have been the best location for a drive lane. There appears to be no reason to judge Paleoindian hunters unable to construct and use fences and corrals similar to those that the archaeological record confirms were employed for communal bison hunting during Archaic and Late Prehistoric times.

The Horner site clearly illustrates the need for archaeologists to have geological expertise. When it was first investigated in the late 1940s under the direction of Glenn Jepsen of Princeton University, researchers contended that the Shoshone River flowed at the site roughly 30 meters above its present level (Schullinger 1951). Three decades later, geological study of the site revealed that the area has changed relatively little during the past 10,000 years, and that therefore the river was at very near its present level during Cody Complex times (Albanese 1987). The different topographies connected with these two interpretations require entirely different bison procurement strategies.

Another 10,000-year-old cold-weather bison kill, containing an estimated 300 animals, was discovered at the Jones-Miller site of Hell Gap age in eastern Colorado in the early 1970s (Stanford 1978). As was the case at the Horner site, the bone counts led the investigator to conclude that the bone bed was the actual kill area. A posthole with several objects, likely to have been ceremonial, around its base and a large hearth nearby with associated patches of yellow and red ochre suggested ritual activity. The hole may have contained a "medicine post," which the shaman climbed to conduct ritual activities before and during the killing of the animals. Relying on historic accounts, Dennis Stanford (1979a) surmised that the bison were driven down an icy path into deep snow that impeded the animals, making it easier for the hunters to kill them with atlatls and darts and thrusting spears. He also suggested that the snow served to freeze the carcasses for extended winter use. The killing event is dramatically portrayed in an artistic rendering that may or may not approach reality (Stanford 1979a: 114–15), but that at least provides a basis for further thinking about Paleoindian bison kills. Without doubt, the posthole and associated artifacts and features strongly support an argument for ritual activity.

As the amount of recoverable site data decreases, the more difficult the task of interpreting that data becomes. One late Paleoindian–age bison kill, the James Allen site in the Laramie Basin of southern Wyoming, is particularly frustrating to those seeking to determine the bison procurement strategy used (Mulloy 1959). The bone bed was found in open country, with a gentle slope to the north; a steeper slope lies directly to the south. Vegetation is limited to grass and small sagebrush. The top of the bone bed lay just below the surface and was covered with windblown deposits. It was discovered because a small part was exposed in a shallow blowout. Covering an area of about 150 square meters, it contained at least fifteen animals believed to represent a single kill.

An arroyo trap is unfeasible at this site, but the location would have been ideal for driving the animals down a steep slope and into a corral as was done at the Boarding School, Piney Creek, and Big Goose Creek sites. The lack of trees in the immediate site vicinity would now make obtaining logs to construct a corral difficult but not impossible, and there is no evidence of postholes from a corral or drive lane. Nevertheless, a corral appears to be the most likely method used, given the site's location and surrounding topography.

Several articulated vertebral columns and leg units led the investigator to conclude that the animals were killed at the location and stripped of flesh, very much as the animals in the bone bed at the Horner site had been treated. Though only fragments of skulls were recovered, the metatarsal and metacarpal measurements indicated that the bones belonged to extinct bison (Berman 1959). Unfortunately, these skeletal elements do not allow the time of year of death to be determined. A bone date of 7,900 years ago fits well with a charcoal date of about 8000 B.P.

on the Frederick (or James Allen) component at the Hell Gap site in southeast Wyoming (Irwin-Williams et al. 1973).

Ecological conditions at the James Allen site and in the southern Laramie Basin area at present are ideal for bison, and they were probably much the same 8,000 years ago. A corral structure seems the most likely procurement method at the James Allen site, but unless there were other kill sites there or in the immediate vicinity, it is difficult to envision constructing a corral for a single, small kill. The absence of other known sites there or nearby tells us little, however; this site was nearly lost through erosion, and it is likely that other bison kill events occurred in the same area but either were buried or eroded away. There is a shallow depression a short distance from the James Allen site where a winter kill of fourteen domestic cattle occurred when they wandered into the feature during a blizzard in 1983. Most of their long bones and ribs are disarticulated, and several skeletal elements including vertebral columns and long bones are partially buried by colluvial and wind-borne deposits. Skulls and pelvic bones are still exposed and rapidly deteriorating. The deposits covering them appear remarkably similar to those that covered the nearby James Allen site bone bed.

The procurement strategy involved at another plains bison kill site is also difficult to understand. The Fletcher site in southernmost Alberta consists of a bison bone bed at the edge of a water hole; it is associated with Scottsbluff-type projectile points. The almost featureless terrain leaves no clue to the procurement strategy involved, if it indeed was a kill site (Forbis 1968). The topography rules out a jump or an arroyo trap, and there was no evidence of a fence that might have formed a corral.

Other bison bone bed sites deny the investigators sufficient evidence to allow a convincing argument for a particular procurement strategy. In addition to sites already mentioned, two other Paleoindian sites, the Hudson-Meng site of Alberta age in western Nebraska (Agenbroad 1978b) and the Plainview site in northwest Texas (Sellards, Evans, and Meade 1947), fall in this category. The Wardell site of Late Prehistoric age in western Wyoming (Frison 1973) might have resulted from driving animals from one direction into a corral similar to that at the Ruby site; or hunters may have driven animals in the opposite direction down a steep slope until they were stopped by a fence similar to the one at the Scoggin site. Yet another possibility is that both procurement strategies were utilized.

It is easy for an archaeologist to become so enamored of the large communal bison kills found on the plains and in the foothills and mountains that he or she overlooks the evidence of smaller and less visible, but still important, strategies that procured bison on a less spectacular scale. The Waterton and Crowsnest valleys in southern Alberta provide good examples of such approaches (Reeves 1978a; J. Driver 1985). The topography in both locations is different from that of the plains, and apparently smaller cultural groups were involved that employed peninsular traps (converging edges of landforms that led into marshes to contain the animals) and ice-block depressions (bowl-shaped, with steep sides). Though very different from the strategies practiced on the plains, these were effective and equally dependent on an acute understanding of bison behavior for their success.

Other tactics mentioned include trapping in snow drifts or lake muds, and possibly driving bison onto lake ice during the winter. Bison on slick ice are at a distinct disadvantage, and are relatively easy to kill and retrieve. Those trapped in mud, in contrast, are difficult both to retrieve and to salvage for edible products. Animals that bury themselves so deeply in snow drifts that they can be approached and killed also present problems of retrieval. It requires a sizable drift to contain a healthy bison; it will struggle as long as possible, and as it works itself more deeply into the snow it makes the hunter's access to it for butchering increasingly difficult. The best time to drive animals into snow drifts is in late winter, when they are weak and tire rapidly. If they can be run for a distance until nearly exhausted and then piled into a snow bank, they are unable to struggle effectively. Under these conditions, they can be approached and killed with a better chance of retrieval.

Many people, sometimes knowingly, cause undue stress to today's wild animal populations by pursuing them during the winter months on skis and snow machines. Late winter and early spring is a critical time, when fat resources are used up and food is scarce. Predators have most success against the old and weak. Wild animals need all their strength to survive until there is green grass—but that is still a month or more away. Prehistoric bison hunters were probably well aware of this cycle; they may have deliberately waited until the animals were weak from the long winter and then used deep snow to immobilize them.

A PRAGMATIC APPROACH TO DRIVING
AND CORRALLING BISON

Successful operation of arroyo traps, sand dune traps, and the many variants of jumps and corrals all depended on the hunters' ability to locate bison herds and eventually maneuver them to a final destination where numbers could be killed, butchered, and processed. We can gain a better understanding of their skill if we consider present-day bison handling. Many bison breeders use cattle-handling equipment for bison, and the difference between the two species, *Bison bison bison* and *Bos taurus,* immediately becomes apparent. The first time I was asked to participate in serious work around bison, I was told to close the corral gate on small groups of animals on their way into a smaller pen, from which they would be forced into a chute that ended in a squeeze gate. The first group, four cows with five- and six-month-old calves, entered the corral at top speed—then suddenly changed their minds and reversed direction; they were out of the corral before I could react fast enough to close and latch the gate. The other workers present had long experience of bison ways and had predicted exactly what would happen; they all had a good laugh at my expense. As is true of most encounters with animals, experience was a far better teacher than any number of verbal instructions given beforehand.

To oversimplify somewhat, there are two ways to handle bison. One is to construct fences and corrals strong enough to withstand the pressure exerted by large numbers of animals; the other is to use less substantial structures and work with fewer animals. Prehistoric hunters tended to take the latter approach. Neither the Ruby nor the Muddy Creek corral was designed to handle large numbers of animals at one time; my guess is that each could hold ten or perhaps fifteen animals without undue stress being placed on wooden components. Anyone who has had to spend time and effort to repair broken fences will seriously consider the consequences before attempting to force too many bison into a corral.

Once inside a corral, bison demonstrate distinctive behavior. The animals along the fence will actually push against the animals crowding into them rather than against the fence. A few animals in a corral are thus unlikely to cause undue strain. But a corral half full of animals can push against a fence with enough force to break off large posts. Because the object of prehistoric communal bison procurement was probably to kill as many animals as possible, hunters were unlikely to be very concerned about high numbers of animals in a corral causing the death or injury of calves and the old and weak. Lending support to this assumption is Thomas Kehoe's description (1967: 76) of a historic Blackfoot bison corral that had sharpened poles extending into the structure, firmly braced on the outside and designed to impale animals that might try to escape.

Bison, unlike cattle, are not domesticated animals and quickly shift

FIGURE 24. Agitated and charging female *Bison bison.* (From Frison 1991a: 21.)

from placidity to wild behavior. When they become agitated, they will test a fence; if they find a hole or a weak spot and can force their head through an opening, they will exert surprising strength and determination as they attempt to enlarge the hole and escape. In the process they often cause themselves serious injury, breaking bones and tearing hide and flesh. I once was faced with a cow coming directly toward me at full tilt; having no time to move out of her path, I crouched down as she jumped over me and totally destroyed part of a fence immediately behind me. It required most of the afternoon to rebuild the fence. The cow in question (figure 24), apparently unhurt, forced her way through two wire fences and was not located again until late the following day, hidden in a patch of thick brush more than 3 kilometers away. Only by maneuvering her into a position where she could mingle with other animals were we able to entice her back into the main corral.

One needs to be alert and able to move quickly to avoid injury while working with bison in corrals and moving them through chutes. An animal confined in a narrow chute can kick viciously and can even break the arm of someone trying to insert a pole behind the animal to prevent it from backing up. However, bison in corrals will usually keep their distance from a person carrying a sizable stick. I believe that a thrusting spear may have been used on occasion on prehistoric bison inside corrals or other features where they were tightly confined.

When bison are placed in mechanically operated squeeze gates designed for domestic cattle, a bison-specific strategy is needed. Human reaction time is fast enough for domestic cows: once one starts through the squeeze gate, the operator can close it, leaving the animal's head protruding through the front and its forward progress stopped by the iron bars in front of each shoulder. We are not always fast enough for bison. Once they decide to enter the squeeze gate, they can move so quickly that even an experienced gate operator is often unable to get it closed in time; and if the bison gets its shoulders past the iron bars, it will never cease struggling until it manages to push the remainder of its body through, in the process often tearing hide and flesh, cracking ribs, and breaking horns and pelvic bones. With narrower shoulders than domestic cattle, bison are more wedge-shaped when viewed from above and can force their way through a squeeze gate that will effectively stop the cow—but not, unfortunately, without injuring themselves.

A common solution in the past was to mount a platform, such as a heavy door, vertically on the rear of a small truck about a half meter off the ground and back it into position about a half meter from the forward opening of the squeeze gate, allowing only enough room for the bison to get its head through the end. The animal then has to pause momentarily and turn its head to one side to attempt to go around the door, giving the squeeze gate operator time to trip the mechanism and immobilize the animal. When the animal is done (with pregnancy testing, brucellosis shots, ear tagging, etc.), the truck is moved ahead a few meters and the animal is allowed to exit; the apparatus is then moved back into its original position.

An astonishingly small space is required by a bison to reverse itself. In a narrow chute, a determined bison will occasionally rear up on its hind legs and use its head and front legs to leverage its body around. When this happens, the animal, and any others behind it in the chute, must be moved back into the holding pen and driven again into the chute—always a more difficult task the second time. Any bison owner will readily admit that it is best to avoid, as much as possible, forcing bison through chutes. This is especially true as calving time approaches, because the associated trauma can cause a cow to abort a calf.

Nearly two decades ago, a local bison breeder, Pete Gardner of

Wheatland, Wyoming, donated for study a cow that failed to have a calf that spring. According to his records, she was more than twenty years old and, since the age of three years, had produced a healthy calf nearly every year. Both horns were missing and she swung her right hind leg sideways as she moved, but she could still travel at surprising speed when closely pursued. As her bones were cleaned of flesh, it became clear that her hip problem was caused by a broken femur whose head had grown solid into the hip socket, with no connection to its opposite end. Even so, her leg muscles were strong enough to swing her right hind leg outward and forward as she traveled. As a result of forcing her way through corral fences and squeeze gates through the years, the dorsal spines of all vertebrae were broken and permanently bent to one side, leaving no evidence of the hump distinctive of all bison. Several ribs had been broken but healed; similar damage was present on both iliac bones. Both horn cores were broken off close to the skull; her jaw had also broken, and though it had healed the resulting misalignment of her teeth caused some problems in eating. An animal with these kinds of disabilities would not have survived long under open range conditions, but her owner gave her supplemental winter rations. Despite all her problems, she managed to stay in good condition and produce enough milk to raise a healthy calf nearly every year.

Gardner was somewhat unusual among the various bison breeders I came to know. At one time, his herd numbered close to 500 and I soon believed his claim that he could recognize each animal by its distinctive personality. He also had a great affection for his animals, kept them well fed, and nursed an animal when it was sick or hurt. He became a true student of bison behavior; by simply observing the animals, whether on the open range in spring, summer, and fall or on winter feed grounds, during good weather or bad, he could immediately detect when one was in distress.

He had one cow that developed into a true troublemaker with totally unpredictable spells of bad behavior. Taking an immediate dislike to a new and unfamiliar pickup truck, the cow charged the front end, smashing the grill and forcing a horn through the radiator, which promptly lost all its coolant. This occurred in a location too far from headquarters for the vehicle to be driven home without damaging the engine, and the cow continued charging the vehicle, causing major damage to its fenders and doors. Finally satisfied, and after keeping Gardner inside the vehicle for more than an hour, the animal made one final charge, then walked away and started grazing contentedly with the remainder of the herd.

Cattle guards are made of metal pipes or I-beams strong enough to support the weight of vehicular traffic, laid parallel to each other at intervals of about 10 centimeters. They are attached to a heavy metal frame too wide for the animals to jump across and placed at a right angle to the roadway on a solid foundation, with an open pit beneath. They are effective barriers to domestic stock and bison, which can neither leap nor safely walk across them; if an animal allows its feet to slip between the beams into the hole below, the result can be a broken leg or legs—often caused, in the case of a bison, by the animal's wild attempts to extricate itself.

One of Gardner's cows wandered away from the main herd one day in late winter and was accidentally left behind when the herd was moved to a new pasture. The cow was desperate to find the herd so it crossed the cattle guard, which had been filled with drifting snow that had become hard-packed. The snow provided enough support to prevent the animal's feet from sinking to the bottom of the underlying pit, so she broke no bones in the process (but lost some hide and hair). Still seeking the herd, the cow, with badly bruised legs but able to walk, started down a roadway. Along the way, a high-spirited horse that had apparently never seen a bison before became frightened enough to charge into and become entangled in a barbed wire fence. The horse was seriously cut and required a trip to the veterinarian. Enraged at his horse's condition, the horse owner grabbed his rifle and killed the bison: the ensuing court action determined that he was liable, since the bison was on a public roadway. He was forced to pay for two animals, because the cow in question was close to calving.

Horses and bison are not always compatible at close quarters. Under certain conditions, particularly when trailing cattle over long distances, ranchers commonly unsaddle the horses and turn them into a corral for the night with the cattle and feed them together. A horse placed with cattle generally tends to adopt a dominant attitude without incident; a bison, however, often resents being pushed aside and swings its head sideways, sometimes goring the horse and causing serious puncture wounds or badly torn hide. An incident in July 2002 illustrates the same behavioral pattern with a human. A man who was walking on a path in Yellowstone National Park passed too close to a grazing bison. The latter swung its head sideways, goring the man in the thigh, and then resumed grazing. The victim was wounded seriously enough to require hospital care.

As long as they are well fed and not bothered, bison are usually content to be contained behind cattle fences; Gardner used this method of control. But some bison, bulls in particular, seem unable to resist the occasional urge to seek out new pastures. When that happens, the animals pay little attention to fences unless they are very strong and very high. Broken fences nearly always result in mixed herds or animals drifting far away from their home territory. Besides the extra work of bringing the bison home, such incidents also often cause friction with the neighbors, particularly if they happen to be cattle raisers.

Fortunately, bison have far fewer problems in calving than do domestic cattle. The trauma associated with catching and holding a bison cow to pull a calf is usually disastrous. The cow resists all efforts at holding her down; and more often than not, once the calf is delivered, she will depart the scene and never return to claim it. Sometimes a calf born late in winter or early spring and out of the normal birthing period will survive and mature slightly ahead of calves born later in the normal birthing period. This animal may breed in her second year rather than her third, before she is sufficiently developed to give birth to a calf without help— a calf that she may then desert without a backward glance. Such late-winter calves would most likely not survive in the wild, but bison breeders try to protect and save them. Above all, bison owners try to prevent the cows from breeding too young to avoid such calving problems.

It is tempting to draw on knowledge of domestic cattle to generalize about bison behavior, but such comparisons serve little purpose: the species have far too many significant differences, as become obvious with close and regular contact with bison. Contemporary observations of the behavior of modern bison on the open range, in corrals, in chutes, and in squeeze gates aid us in understanding prehistoric bison corrals and their associated features. They also serve to make us critical of many commonly accepted reconstructions of prehistoric hunting practices derived from questionable sources by those unfamiliar with animal behavior. A caution to those investigating bison kill sites: if there is a reasonably accurate reconstruction of the site area topography at the time of the site's operation, and the proposed strategy for trapping the animals is not consistent with known bison behavior, it is time to seriously reconsider the interpretation of the data.

It is also worth noting that the food preferences of bison and domestic cattle differ. While I was driving down a road in western Kansas that separated a large bison pasture on one side from a cow pasture, my passenger pointed out that in the bison pasture the yucca plants were missing, while in the cow pasture they constituted a significant part of the vegetation. Neither of us could explain this phenomenon, which Pete Gardner told me he had also observed among his own animals. As he closely watched his bison through field glasses, he saw them approach a yucca plant; ignoring the sharp spines and using incisors and upper gums, they grasped the centrally located, more succulent spines, pulled them out, and ate them, thereby killing the plant. This was clearly a choice on the part of the bison, since the pasture had more than adequate grass. Domestic cattle have to be desperate for food before they will attempt to consume yucca spines.

Both for economic reasons and also because they are viewed with affection by many people, such as Pete Gardner, bison have become more popular over the past two decades and their numbers have increased dramatically. Bison are able to utilize grasslands better than domestic cattle, and their meat, while comparable in taste to beef, is believed to be healthier. The many recent improvements in bison handling include electrically and hydraulically operated squeeze gates with metal cages on the front to prevent an animal from advancing too far into the mechanism and injuring itself while attempting to force its way through—a welcome replacement for the solid door on the rear of a truck. Such innovations make bison handling easier and significantly reduce the animals' trauma and injuries.

In August 2000, this heightened interest in bison brought together more than 1,200 individuals from thirteen countries to a conference in Edmonton, Alberta. Those attending included bison producers, specialists in various aspects of bison management and research, and others interested in or just curious about bison. One researcher who has probed deeply into the psychology of bison handling points out that they present "a unique management situation" and need a "proactive handler who stops to check the facility for distractions such as shadows, contrasts, litter and dangling objects like string and chains before working animals." She finally notes, "Good handlers can work animals in poor facilities, but poor handlers cannot calmly work animals in the best of facilities" (Lanier 2001: 265–68). I agree wholeheartedly with these observations and am convinced that they were probably as well known and usefully applied by prehistoric bison handlers as by those of the present day.

WEAPONRY, TOOLS, AND BISON HUNTING

Some of my own most frustrating moments in hunting have occurred at the end of a successful stalk when the targeted animal disappears in the distance after my weapons have failed. This failure can be the result of inadequate equipment, improper maintenance, or simply inferior materials. A hunter soon learns the sad consequences of neglecting these important aspects of his subsistence strategy. What I see in the archaeological record is that through necessity and by trial and error, prehistoric bison hunters developed efficient weaponry systems. To that end, they attempted to procure the best tool stone available. There is ample evidence of their efforts. Early-nineteenth-century European settlers on the plains encountered many large tool stone quarries; and having been raised to believe that Native Americans were inherently lazy, they attributed them to Spaniards digging for gold. The Spanish Digging in eastern Wyoming is one well-known example (see Holmes 1919), and the Spanish Point Stone Quarry in the Big Horn Mountains in northern Wyoming is another (see Frison 1991b: 291–92). Only during the later periods of prehistoric bison hunting were perishable parts of weaponry assemblages preserved, which were found not in actual kill sites but in dry caves and rock shelters (see Frison 1965, 1968a). Although limited in number, they are about the only counterparts considered reliable enough for reconstructions of older weaponry systems.

Paleoindian bison hunters were active for approximately 3,000 years, from about 11,200 to about 8,000 years ago, and they used several different variants of a functional design for their flaked stone weaponry components. The superior design of the Clovis projectile point has already been discussed in the context of mammoth and elephant killing, but all later Paleoindian projectile points (figures 25a–e) can be viewed as variants of the same design: it incorporates a sharp point for penetrating the hide, sharp blade edges to follow the initial puncture and cut a hole large enough to allow the remainder of the point and foreshaft to enter the animal, and dulled lateral edges toward the base to prevent the cutting of sinew bindings under stress. Paleoindian knappers were skilled in reducing stone to the desired shape—so skilled, in fact, that many of their products may be works of art as well as of functional utility. One accomplished lithic technologist, Bruce Bradley, has suggested that the act of fluting Folsom points could have been part of a prehunt ritual (Bradley 1991: 373–79).

Other finds suggest that ritual significance was expressed in weaponry. Examples include an Allen point from the Norton bison kill in western Kansas (Hofman and Graham 1998: 106), a miniature Hell Gap point from the Jones-Miller site in eastern Colorado (Stanford 1978: 97; 1979a: 119), a Goshen point from the Mill Iron site in eastern Montana

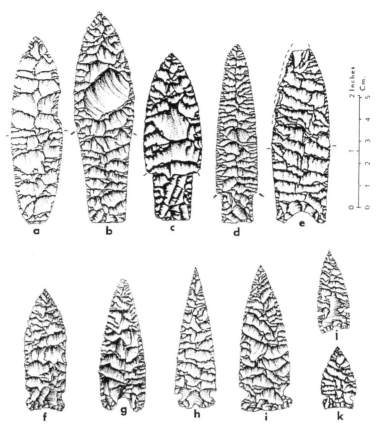

FIGURE 25. Selected projectile point types from bison kill sites: *a*, Agate Basin point; *b*, Hell Gap point; *c*, Alberta point; *d*, Eden point; *e*, James Allen point; *f*, early side-notched point; *g*, McKean lanceolate point; *h*, Yonkee point; *i*, Besant point; *j*, *k*, Late Prehistoric side-notched points. (From Frison 1978: 158, 180, 186, 204, 225; 1974: 74; 1991b: 64; Frison, Wilson, & Wilson 1976: 43.)

(Bradley and Frison 1996: fig. 4.3d), a Folsom point from the Cooper site in Oklahoma (Bement 1999: fig. 43FF.), and a miniature point from the Muddy Creek bison corral (Frison 1991b: fig. 2.60d). All demonstrate truly exceptional and even artistic craftsmanship in their manufacture, but their morphology raises questions about their functional value for killing large animals. They stand out from the remainder of the associated assemblages; they may have been deliberately placed in kill sites for ritual purposes and never used as weapons.

Some form of a throwing stick or atlatl is believed to have been the means of delivering most projectiles before the introduction of the bow and arrow (at the beginning of Late Prehistoric times, between about 1,500 and 2,000 years ago). Both thrusting and throwing spears may have been used in certain contexts (for an artist's rendition, see Stanford 1979a: 114–15). However, evidence for atlatl use in Paleoindian times is extremely limited. One object believed to be an atlatl hook made from a section of deer antler came from the bone bed of the 9,000-year-old Jurgens site in Colorado (Wheat 1979: 135–36). The porous interior of a section of deer antler was removed and a hook was formed on one side of the outside shell. The distal end of the atlatl, presumably of wood, would have extended through the central hole and then been securely anchored. Joe Ben Wheat has also suggested that internal cores of bison lower molars might have been used as atlatl hooks (1979: 95, figs. 46b, c), but this seems doubtful: my experiments with their use as atlatl hooks proved them to be too brittle to withstand the pressures exerted.

The tip of an elk antler tine from the 10,000-year-old Hell Gap–age bison bone bed at the Agate Basin site is believed to be an atlatl hook (Frison and Craig 1982: 164). If so, it would have been designed differently than the one of deer antler described above. The large end of the elk antler tip would have been solidly secured in a hole in the distal end of the atlatl shaft and the sharpened tip would have protruded beyond the atlatl shaft and formed the spur. My experiments utilizing an elk antler tip as a hook in this manner proved successful.

The 11,000-year-old bison bone bed at the Goshen-age Mill Iron site in eastern Montana produced a section of mammoth rib with a cone-shaped hole drilled in one end and a transverse break on the opposite end that is believed to be the distal end of an atlatl main shaft (Bradley and Frison 1996: 67). It would have been attached to the wooden part of the shaft. The shape and size of the hole would readily accommodate the largest of the wooden foreshafts found in Late Archaic sites. A shaft with a tapered end that fits into a matching conical hole in another shaft is ideal for withstanding the forces of impact when it is used as a weapon, but not the pressures exerted on it during use as a tool. There is no known evidence to indicate that mammoths were present in Goshen times, but the Clovis mammoth kills were recent enough that their skeletal remains might still have been available. However, I readily admit that these objects are far from providing conclusive evidence of Paleoindian-age atlatl components.

My first assumption about projectiles recovered in a kill site bone bed is that they were used in killing the animals. A further assumption is that they were recognized by the hunters as having the potential to successfully conclude their hunting episodes. Some projectile points in kill sites appear to meet these criteria, but others demonstrate modifications believed to have resulted from their use after the kill, probably as butchering tools. Still others demonstrate reshaping, presumably of damaged points, to restore them to a condition that will enable them to be used once more for killing animals. However, these reshaped points rarely approach the technological expertise expressed in their original manufacture.

Variants of the flaked stone components of Paleoindian weaponry identify the presently known cultural complexes associated with bison kill sites (see figures 25a–e). All can be securely hafted in a notched foreshaft with a minimum amount of sinew. If a foreshaft and mainshaft were used instead of a single long shaft, the taper on the foreshaft should closely match the conical hole in the distal end of the mainshaft. Joe Ben Wheat (1979: fig. 46e), however, has suggested that a mainshaft terminated with a conical tapered distal end that was inserted into a conical hole in a large foreshaft. Whatever the actual configuration of mainshafts and foreshafts, when assembled the elements needed to be in perfect alignment: any deviation reduces the amount of thrust to the base of the projectile and increases the possibility of poor penetration or damage to the equipment. At close range, up to 20 meters, fletching on the mainshaft improved accuracy; for more distant targets it was definitely necessary.

I have been unable to establish any one Paleoindian-type projectile as significantly more lethal than another. However, as is true today, individual hunters had definite preferences, as indicated both by the range of variation in a single type in individual bison kill site assemblages and by the different types used in different cultural complexes over time. My own preference is for the Agate Basin point (figure 25a); its thick, lenticular cross section provides structural strength and enough weight to ensure penetration. Its tapering base and sinew wrapping absorb the shock of impact, and transverse breaks usually result in sections that can be retrieved and easily reworked for further use.

I have been unable to understand the popularity of the Hell Gap projectile point (figure 25b), which apparently evolved directly out of the Agate Basin point: it adds a shoulder, which transferred much of the shock of impact there rather than to a tapered base. I have not found it to penetrate as deeply as does the Agate Basin point, but it cuts a larger hole on impact and may have had other advantages I have not detected. The Alberta projectile (figure 25c) introduced a shouldered and stemmed point, directing the shock of impact to both the shoulder and the base.

Refinements of the Alberta point resulted in the Eden and Scottsbluff points that are specific to the Cody Complex (figure 25d); my second choice of a projectile for its effectiveness on bison is the stemmed Eden type with either lenticular or diamond-shaped cross section. The James Allen point (figure 25e), thin in cross section and with an expanding base, penetrates well but snaps easily on impact. It apparently functioned adequately, however, given its success at the James Allen bison kill site in southern Wyoming (Mulloy 1959), even though the points recovered in the bone bed reveal an unusual amount of impact fracturing and shattering as compared to other Paleoindian point types (with the possible exception of Folsom).

Projectile points from the Hawken bison kill site (Frison, Wilson, and Wilson 1976) demonstrate a new idea in projectile points used to kill bison. Sometime within the 1,400 or so years between the use of the Hawken site and the presently known final Paleoindian bison kills, side notches were added to the lanceolate Paleoindian projectile points (figure 25f). Notches are believed to have simplified the hafting process without reducing the point's deadliness. But side notching apparently had one negative effect: it increased the tendency of stone points to break across their notches. Lanceolate points (figure 25g) reappeared in Middle Plains Archaic times in the McKean complex about 5,000 years ago, but some notched varieties were also in use, as the Scoggin bison kill in central Wyoming described above clearly demonstrated (Lobdell 1973). Various styles of notching lasted until the end of the use of stone projectile points to kill bison.

Late Archaic bison kills produced some impressive projectile point assemblages. One group of arroyo bison kill sites along the Powder River and its tributaries in northern Wyoming and southern Montana is known as the Yonkee bison kills, named after the owners of the first site of this group that was systematically excavated, the Yonkee site (see Bentzen 1962). It was first believed to be part of the McKean complex of Middle Plains Archaic age, because charcoal collected during the 1962 investigation was dated to about 4,500 years ago; a more recent investigation with better stratigraphic control put the age of the site at about 3,100 to 2,700 years (T. E. Roll, letter to author, 1988). A male bison skull collected near the site, but not in the bone bed, during the 1962 investigations was first postulated to be an intermediate-size animal with horn core spread larger than the present-day bison, but a subsequent assessment disproved this conjecture (M. Wilson 1975: 216). All of the Yonkee arroyo kills and one bison jump, the Kobold site in south-central

Montana (Frison 1970b), which has a Yonkee component, are now regarded as Late Archaic in age. At three Yonkee bison kills in arroyos—Yonkee (Bentzen 1962), Buffalo Creek (Miller 1976), and Powder River (Frison 1968b)—butchering at the sites consisted of stripping flesh and leaving skeletal articulated elements, much as the bison at the Horner site were treated (as described above). Favored targets at the three sites were rib cages, although two projectiles were lodged together in the neck of one animal at the Buffalo Creek site.

Projectile points in the Yonkee sites are well designed for killing bison; they are long and narrow with sharp points and have either side or corner notches (figure 25h). Base notches or indentations could have been intended for better securing the point in a notched foreshaft and preventing sideways movement during impact. Most of these projectile points were made of a metamorphosed shale formed when clay beds were fused by intense heat from coal beds that burned in the Powder River Basin in northern Wyoming and southern Montana during the late Pliocene or early Pleistocene. This material occurs in several colors and qualities; the best equals the quality of the best cherts and quartzites for projectile points.

Some insight into the effectiveness of a Yonkee projectile point (e.g., figure 25h) on bison came from the Buffalo Creek site (Miller 1976). In one case, the projectile point penetrated the bison's hide close to the top of the back, continued through the top of the rib cage cavity, and then penetrated about 14 millimeters into the base of the fifth thoracic vertebra's center (figure 26a). Apparently the hunter was slightly above the animal—as one might expect, since the bison are believed to have been trapped at the headcut of a steep-sided arroyo and the hunters would have been shooting from the arroyo bank above. It was not the best of shots (figure 26b), but it clearly indicates the penetrating qualities of a well-designed projectile point.

The Besant groups were probably the last Plains bison hunters to use the atlatl and dart. The two Besant bison corral sites described above, Ruby and Muddy Creek, yielded large numbers of well-crafted side-notched projectiles with sharp points that were manufactured of the highest-quality cherts and quartzites known to have been available (figure 25i). It would be difficult to choose between Yonkee and Besant points if one wished to inflict lethal wounds on bison.

The bow and arrow eventually won out over the atlatl and dart. Bison kill sites from about the past 1,500 years reveal a wide array of different styles of the smaller projectile points assumed to indicate a change

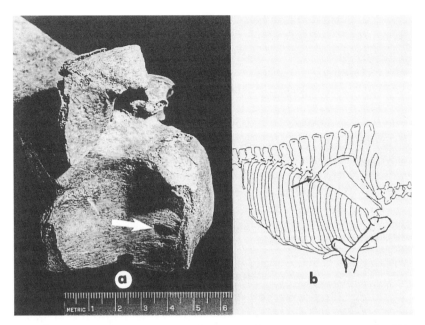

FIGURE 26. *a*, Penetration of a Yonkee projectile point 14.4 mm into the fifth thoracic vertebra of a mature female bison (from Miller 1976: 52) and *b*, path of entry of the point into the animal.

to the bow. Prehistoric hunters may have been much like those of today in that they preferred a smaller projectile (figures 25j, k) with greater velocity, longer range, and better accuracy. Smaller points are easier to manufacture and consume less raw material than the larger dart points. Yet whatever disadvantages the atlatl and dart had, it was probably difficult for some hunters to change from the older weaponry system to the new, since the atlatl and dart in the hands of an expert is a highly effective weapon.

AFTER THE ANIMALS HAVE BEEN KILLED

It always comes as a surprise how quickly carnivores and scavengers, both animals and birds, detect a dead carcass. Usually before a dead animal is field dressed, a magpie, raven, or jay appears on the scene; during warm weather, flies arrive even sooner. The smell emanating from a bison kill of several animals must soon have attracted a varied audience, as is nearly always confirmed by the presence of their skeletal remains. Those remains raise the possibility that animal kill sites may have served a secondary purpose as lures to attract other animals and birds. Indeed, it seems possible to me that some of the projectile points found in animal kills that are of questionable utility for killing large animals such as bison may be present because they were used to dispatch the carnivores and scavengers attracted by the odor of decomposing animals killed earlier.

In the hunting process there is always an element of uncertainty until the animal draws its last breath. More than once, I have observed a novice hunter view his quarry stretched out on the ground, lay down his firearm in the belief that the animal has expired, and approach it to field dress it—only to have it suddenly jump to its feet and, at top speed, disappear into the brush or over the first rise, leaving the hunter too surprised to retrieve his weapon and fire another shot. Sometimes the escape is the result of the animal's last shot of adrenaline, in which case the animal will usually be found a short distance away; sometimes the animal flees because it was only temporarily stunned and suffers few, if any, negative aftereffects. I have witnessed this numerous times in the case of a deer or elk that was stunned for several minutes by the shock of a bullet hitting a main antler beam.

However, once the death of an animal is confirmed, the attitude of the hunter changes. His main objective has been reached, but he still has another serious responsibility. Sometimes the loss of valuable meat products is unavoidable or accidental, but most often it results from a hunter's neglect—and few errors could more quickly puncture the joy at success in hunting. Butchering an animal properly is the first step to maximizing the amount and quality of the meat products. The process is simpler if the weather is cool, and if the animal is in the open and on a level piece of ground covered with snow or heavy grass; much more effort is needed to butcher the animal and save the meat in hot weather and when the animal dies in heavy underbrush or in the bottom of a deep arroyo. In the case of a mature bison, which will usually weigh about half a ton, the mere contemplation of the work to come rapidly dissipates the euphoria of a successful hunt. One person can quickly field dress and transport an animal the size of an average deer, pronghorn, or mountain sheep, but not a bison, which must be cut into pieces small enough to handle. The help of one or more persons to hold carcass parts in the optimum positions for skinning, stripping flesh, and disarticulating joints is most welcome when a hunter is butchering a large animal such as a bison.

The hide and flesh of any animal can be removed with a small tool made of stone or metal, provided it has a sharp edge that can stay sharp through the entire process. I have demonstrated this numerous times by skinning all or parts of bison that died from various causes; usually these were young, weak, and older animals, killed by overcrowding and trampling in chutes or corrals (see Frison 1991b: 303). However, without exception, these were animals that had been dead for some time. If it was winter, they were usually frozen; if the weather was warm, they were often so badly deteriorated that I could not endure the stench. Knowing that a freshly killed animal is easier to skin than one that has been dead long enough for rigor mortis to occur, I was constantly on the alert for the opportunity to experiment on one of the former.

In the winter of 1975, the manager of a state park in northern Wyoming informed me that he would provide a bison from a small state herd that I could butcher with stone tools, provided a meat inspector was on hand to guarantee the sanitary condition of the meat, which was to be used at a convention banquet. The animals were on open range with no supplemental feeding: earlier in the winter would have been a better time to kill an animal because it would have been in better condition. The animal selected, a mature female that had lost her calf the summer before, was killed and placed on a large canvas, even though the several inches of fresh snow on the ground would have prevented dirt from contaminating the meat.

The skinning and stripping of the flesh progressed as planned with one side of the animal, but by the time the animal was flipped over and skinning began on the opposite side, I would have gladly changed over to a metal knife. However, a large crowd of curious onlookers, many with cameras, had assembled, and I knew that to save face the job would have to be finished using stone tools. Fortunately, one of my students who was along helped me skin and strip the second side of the animal. Even so, we were both tired of butchering with stone tools by the time the task was done.

When the animal was completely butchered, the meat inspector admitted it was as clean a job as anyone could have accomplished with a metal knife. He even managed to slip me a piece of loin, which turned out to be as tender and tasty as any beef. Minus the head and part of the neck, to which someone had prior claim, the carcass produced 368 pounds of meat—reasonably close to the generally accepted average of 400 pounds (182 kg) for female bison claimed by Theodore White (1953a: 397). Had the animal been in better condition, it would undoubtedly have yielded several more pounds of meat.

Once again, experience proved a better teacher than secondhand observations and advice from those unfamiliar with the actual use of stone tools. Stone skinning tools used in my early experiments were copied from those recovered in archaeological contexts and commonly identified as biface knives. They did not function as well as expected, because we were copying tools resharpened so often that they had been worn out and discarded. Large, thin percussion flakes with a unilateral retouch on a low-angle cutting edge (30 to 35 degrees), resharpened as they became dull, functioned much more efficiently. However, in my opinion, stone knives are a distant second to a good metal knife for cutting bison hide, stripping flesh, and disarticulating joints. Metal holds an edge longer and is easier to resharpen than stone. Using a stone knife requires different muscles than using a metal knife, and blood and fat make a stone knife difficult to grasp. To be sure, I realize that more experience than can be acquired from butchering a single bison is needed to claim much in the way of expertise. After one has butchered several animals, the muscles become conditioned, and choosing optimal tool size and preparing and maintaining a working edge become second nature. But unless the hunter is out to prove a point, I have yet to see anyone choose a stone butchering tool over a metal one.

One accustomed to metal knives who then tries a stone knife will find the habit of exerting sideways pressure when cutting difficult to break. A handle of some sort aids in the use of a stone knife, but to attach a handle firmly enough to withstand the pressures of cutting heavy hide requires bindings that cover a large portion of the tool. Sinew bindings are difficult to maintain, because any movement in the haft brings them into contact with body fluids that cause them to stretch and lose their secure hold. In addition, unless the part of a stone knife that comes into direct contact with the hand is deliberately dulled, it can inflict almost as much damage to the butcher as the cutting edge causes to the animal. Yet with continued practice, stone tools can be made to function with surprising ease and with the efficiency demonstrated by their more than 11,000 years of use in butchering prehistoric bison and other animals.

Because of the sheer number of bison kill sites and of the varied ways they were formed, it is inevitable that an investigator is occasionally led astray. Some of these errors are rooted in a single source, one that I remember vividly: the drought years of 1934 and 1935, when dire conditions in the western states led the Agricultural Adjustment Administration to purchase large numbers of cattle from despairing ranchers (see Petrie 1934–35). Deep pits, dug with horse and mule power, were filled with the carcasses of animals unfit for butchering. Flood waters from heavy rains and spring snow melt along with ground disturbance by construction activity occasionally reveal one of these cattle kills. On two separate occasions, I was informed of bones eroding out of arroyos that at first appeared to be typical prehistoric bison kills. However, to the disappointment of my informants, the carcasses proved to belong to domestic cattle. In one instance, the identification was quickly confirmed—skulls had bullet holes between the eyes. At a second site, flash flood waters scattered disarticulated skeletal elements for more than 100 meters along an arroyo bottom where, once exposed, they decomposed rapidly. In this case, identification was more difficult and rested on careful comparisons between the skeletal parts of bison and domestic cattle.

MISCONCEPTIONS OF BISON BEHAVIOR

In the absence of eyewitness accounts, we rely on artists' reconstructions of bison jumps that commonly depict hunters shouting and waving a blanket as running animals leap into space and then fall to their death at the bottom of a cliff (figure 27). Forcing bison over a precipice in this manner requires a densely packed herd of animals moving so fast that the main bulk of the herd will push the leaders over the edge—but this is not how bison behave. Their ability to change course, stop, or execute a U-turn while running has been well documented. Every time I am around bison or look at any prehistoric kill site, I recall vividly my earlier encounters with bison, both afoot and on horseback, and I consider how similar knowledge of their behavior might have affected the strategies of prehistoric hunters.

Artists and others unfamiliar with bison behavior and with hunting also often depict bison being killed in bogs. In reality, bison are attracted to bogs and swamps, which provide lush grasses for grazing and offer some relief from summertime swarms of gnats and flies; but only very old, weak, or crippled animals are unable to extract themselves at will. The better strategy is for the hunter to drive the animal out of the swamp and kill it on solid ground. Furthermore, he knows that an animal unable to extract itself from a boggy area is of inferior quality; a competent hunter would be too proud to kill such a one and expect the rest of the group to accept it as food. The difficulty of butchering a bogged animal and of

FIGURE 27. Artist's erroneous conception of jumping bison. (From Frison
1991a: 17; taken originally from Gladwin 1947.)

retrieving the meat in edible condition is another reason to drive the animal out of the bog before killing it. I learned this lesson myself at an early
age when I killed a bull elk grazing in a bog. Nearly half of the meat was
unfit to eat by the time it was on solid ground and the contaminated parts
were trimmed and discarded.

In a different and even worse case, an animal is trapped in a quick-
sand pocket or in deep mud in ponds or slow, meandering streams, from
which extraction is almost impossible without a stout rope and one or
more strong horses (and thus entirely beyond the ability of prehistoric
hunters). Domestic cattle behave just as bison do around swamps and
bogs (see Frison 1991a: 20).

Another often-proposed bison procurement strategy that needs careful reconsideration is the use of fire drives. Controlling the movements
of a herd of bison requires split-second decisions and reactions by the
drivers, but a wildfire cannot be micromanaged. Wind direction, wind
velocity, humidity, vegetation cover, topography, and other parameters
determine the progress of range fires, and it is extremely unlikely that
these conditions could be harnessed to move a herd of bison from one
designated location to another at any given time. However, burning tall
dry grass does encourage the rapid growth of new shoots that attract

grazing and browsing animals, thereby making it far easier to predict where they might be expected to congregate.

The first bison migrated across the Bering Strait into North America, perhaps more than a million years ago. The earliest autochthonous North American bison was *Bison latifrons,* which was probably the ancestor of later North American bison. The bison of the late Pleistocene and early Holocene, now referred to as *Bison antiquus* and *Bison occidentalis,* were larger than modern bison—*Bison bison bison* (the plains bison) and *Bison bison athabascae* (the northern or wood bison)—which reached their present size by at least 5,000 years ago.

The first unequivocal North American bison hunters were Clovis, at just over 11,000 years ago; there followed a 3,000-year sequence of Paleoindian bison hunters with highly efficient weaponry who were able to carry out communal kills, utilizing both natural and artificial features to gain advantages over the animals. After about 8,000 years ago, adverse climatic conditions seem to have steadily decreased the numbers of both hunters and bison; those who remained sought more favorable areas both within the plains and mountains and on the eastern periphery of the plains. The killing of bison diminished and probably ceased entirely over much of the plains until somewhere between 5,000 and 6,000 years ago, when improved climatic conditions once again allowed bison to return to the plains and human hunters to return to bison hunting as a major part of their subsistence strategy.

Though we cannot know whether the modern and the extinct bison had similar patterns of behavior, both were taken in arroyo traps and various kinds of corrals. The major new innovation was apparently the buffalo jump, which took many forms. Bison were killed at the Head-Smashed-In jump in Alberta beginning at least 5,500 years ago and possibly even earlier. Extensive drive line systems are associated with jump locations, and their purpose may have been ritual as well as functional. Evidence of religious activity is well preserved at one corral site, and such activity can most likely be generalized to almost all prehistoric communal bison-hunting efforts. Constructing corrals, fences, and other structures was a labor-intensive activity, which would have been undertaken only if enough animals were present to justify the effort. Most presently known Paleoindian communal bison kills occurred in cold weather, when it was possible to freeze surpluses for temporary storage. Subsequent communal kills appear more common toward late summer and early fall, when meat surpluses could be dried for short-term storage.

The continual geological activity at the sites of most bison kills, and those of Paleoindian age in particular, makes expertise in geology essential if the investigator is to correctly identify the landforms utilized. Different landforms required different procurement strategies; they determined the numbers of hunters needed and how the animals were to be driven and trapped. Arroyo traps were of great importance in bison procurement, but they are inherently unstable and thus much of the evidence has unfortunately been destroyed. Sand dunes are even more unstable, and satisfactory evidence of their use as animal traps is often very difficult to detect. The bison jump is the most visible of the past bison procurement features because the drive lines and the high cliffs with bone beds at their bottom have survived. When corrals and traps were used, the hunters had to kill the animals individually; with high cliffs, the primary means of death was the bison jump itself. Because they left large quantities of bone, communal kills are the most obvious and attractive subjects for research into bison procurement in the past; much less effort has been spent investigating the smaller and less visible day-to-day hunting events that involved fewer animals and hunters working alone or in small groups. Yet the latter procurement strategies undoubtedly resulted in killing of many more animals overall than the current total recorded in the large communal kills.

As the Native American tribes on the plains acquired horses, their entire bison procurement strategy changed. Bison jumping was no longer necessary. It was more productive and prestigious to run down the animals and kill them from horseback. Before the advent of horses, the bison had the upper hand; now, the human hunters had greater control. To confirm the magnitude of the change, one need only compare the difference between handling bison on the open range on horseback and on foot. The likelihood of actually experiencing this comparison is rapidly diminishing, however, as all-terrain vehicles are rapidly replacing the saddle horse; in the process, our observations on bison behavior are made under conditions removed yet another step from those present in prehistoric and early historic times.

CHAPTER 5

The North American Pronghorn

A MAMMAL ESPECIALLY ADAPTED
TO THE NORTH AMERICAN PLAINS

Antilocapra americana, the species often referred to as *antelope* that lives at present on the plains of North America, is not a true antelope, a designation properly reserved for species found in Africa. To avoid confusion, the preferred common name for the North American species is *pronghorn.* Because the inclusion of *capra* in the nomenclature suggests some relationship to goats, they are frequently called "stinking goats"— another misnomer. While it is true that during the rutting season mature males acquire a strong smell (which detracts from their desirability as food), the pronghorn fossil record fails to confirm any genetic connection to the goat.

Another bit of folklore is that pronghorn are better eating in the spring than at other times of the year. However, just before World War II, there were spring hunting seasons for pronghorn in parts of Wyoming. I was unable to detect any superiority in flavor in pronghorn killed in spring rather than fall, as long as rutting males were avoided. Present-day hunters are not alone in commenting on the undesirable taste of animals during the rut; one of Gilbert Wilson's Hidatsa informants made the same observation about all male animals (G. Wilson 1924: 230).

There is another likely reason why we associate pronghorn with an undesirable odor. Their gestation period lasts from 230 to 250 days, about two months longer than that of mule deer *(Odocoileus hemionus)*. However, birthing periods and hunting seasons for both species largely coincide, the former peaking about the end of May and the latter beginning about the end of September. Consequently, the pronghorn hunting season is in full swing during the height of the rut, while the mule deer hunting season is nearly over in most areas when their breeding season starts (usually toward the end of November). At that time, mature male mule deer acquire a strong odor every bit as objectionable as that of rutting pronghorn. I believe it is largely the timing of current hunting seasons that has resulted in pronghorn earning the "stinking goat" label.

Pronghorn are unique in that they are the only plains mammalian species with horns that are shed annually. Only the horn sheaths loosen and are shed; the horn cores remain intact. The dark color and small size of pronghorn's shed horn sheaths make them much less visible than the shed antlers of deer and elk, and therefore they go largely unnoticed. Pronghorn are a true symbol of the North American Plains, but overhunting brought them dangerously close to extinction at the beginning of the twentieth century. Once they were protected, however, their numbers rose rapidly enough to allow some limited hunting. Their recovery was enhanced by their ability to adapt to a wide range of ecological conditions and by their productivity—adult females usually produce twin offspring. Extreme winter conditions still drastically reduce pronghorn numbers locally, but they rapidly recover and are at present a common sight on much of the open plains and badlands of the western United States.

The fossil record of antilocaprids in North America extends back into the Miocene with forms characterized by forked permanent horns, but we do not know whether they had horn sheaths that were shed annually. The fossil evidence leading from these to *Antilocapra americana* is unclear, mostly because of a lack of intermediate forms, apparently attributable to the poor preservation of fossil remains (Walker 2000: 14–16). Unlike with bison and mountain sheep, there is no evidence in archaeological sites from Clovis to historic times that today's pronghorn are smaller than their late Pleistocene–early Holocene ancestors (Chorn, Frase, and Frailey 1988). In areas such as the Great Basin that were less favorable to bison, pronghorn were an important source of meat for prehistoric human groups, possibly even as important as bison. Because their behavior and ecology differ markedly from those of other mammals of similar or slightly larger size such as deer and mountain sheep, the strategies for hunting them are also different. However, once their behavior is understood, pronghorn are relatively easy to obtain either by individuals, by small groups of hunters, or by communal hunts involving long drive lines, fenced enclosures, and large numbers of people. The archaeological record confirms that these pronghorn procurement strategies have been carried out successfully for many thousands of years.

Pronghorn ordinarily live about seven to eight years, although a series of years with good summer grass followed by mild, open winters can extend this life expectancy by two or three years. As is true of all grazing and browsing ungulates, tooth condition largely determines life expectancy. Poor feed conditions increase wear on both molar and incisor teeth, a strain that inevitably reduces the animals' life span. The growth of food plants during dry summers is restricted, resulting in shorter grasses and browse plants, while in droughts more abrasive particulate matter settles on plants and increases tooth wear. A dry summer with restricted plant growth followed by a long, harsh winter is very hard on pronghorn. These conditions affect all grazing and browsing wild animals but most acutely the very young, the very old, and the weak, many of whom are unable to survive until spring.

Pronghorn and deer have major behavioral differences, but they often share a large part of the same ecological area, although each exploits it differently. Because pronghorn lack accessory carpals (dew claws) on their feet, in mud, soft dirt, or snow their tracks are easy to differentiate from those of deer. Sure and obvious signs of male pronghorn are conspicuous territorial markers made by pawing in the dirt with their front feet. These markers are usually a meter or so in length and half that in width, visible as spots free of vegetation along trails and arroyo banks. Occasionally pronghorn will paw at a bone exposed at the surface by erosion. I know of two instances when they pawed and scattered prehistoric Native American burials, apparently attracted by the white color of bleached bone.

Pronghorn have high-crowned teeth in contrast to the short-crowned, browsing teeth of deer and elk. They can survive on sagebrush and indeed are often forced to rely on it and other shrubs during the winter months—especially in the northern part of their range, where they frequently must paw through deep snow for food. Well-intentioned efforts to feed hay to pronghorn in danger of starvation from deep snow and intense cold are usually futile: their digestive systems (unlike those of deer and elk) are unable to cope with this kind of feed. During the warm months, they are attracted to lush and tender feed found in irrigated fields and green pastures.

Unlike deer, pronghorn are relatively visible throughout the day. They prefer open, treeless country and like to select a spot to lie on from which they can view the surrounding country. They have excellent eyesight over long distances and rely on their speed to elude predators; they have been clocked at speeds exceeding 60 miles per hour. In the 1930s, I saw a man traveling with a pet African cheetah turn it loose to pursue a pronghorn. The cheetah was rapidly gaining on and about to overtake the pronghorn, a young female, until she sailed over a deep ravine that the big cat refused to negotiate. The cheetah's owner rapidly retrieved it and immediately departed the area to avoid the possibility of running afoul of local law enforcement. This one incident unfortunately did little to settle an ongoing argument as to which is the fleetest animal, the North American pronghorn or the African cheetah.

Pronghorn can jump relatively long distances at high speeds, though they approach fences quite differently than deer. Instead of hopping over, they become airborne some distance before they reach a fence and sail over it. Females and fawns sometimes slide under the bottom wire of a barbed wire fence at top running speed, usually leaving wads of hair behind. As they hug the ground, they barely lose momentum. I have not seen large males do this, probably because their horns would catch in the bottom fence wire. Unfortunately, however, weak animals sometimes fail to clear the fences, catching their hind legs between the top and second wire. Unable to extricate themselves, they suffer a long and lingering death. Pronghorn usually prefer to crawl under or through fences rather than jump them. As a consequence, sheep-tight fences without occasional openings can disrupt their migration patterns; under conditions of deep snow and intense cold, the result is often mass deaths. In addition, there are well-documented accounts of large herds moving to the leeward in near whiteout conditions during intense winter blizzards and perishing as they plunged over steep precipices (see Lubinski 1997: 187–206).

During the past half century, pronghorn have adapted surprisingly well to fences. In the late 1940s, I accidentally cornered a herd of about a dozen pronghorn at a location in the Powder River Basin in eastern Wyoming where a single strand of barbed wire from an abandoned fence line happened to lie in plain sight on top of thick sagebrush about a meter above the ground. It formed an open-ended oval of approximately 30 by 40 meters. When the animals reached the closed end of the loop, they refused either to jump or to crawl under the wire. Further crowding of the animals caused them to double back and run by me at top speed. Faced with the same situation today, the animals, now familiar with wire fences, would almost certainly go under the wire to escape.

FIGURE 28. Artist's conception of golden eagles attacking pronghorn. (From Maycock 1980: 5.)

Aggregations of pronghorn are reliable indicators of the time of year. Animals are usually in good condition in the fall of the year following the rut, and then they begin to congregate in herds. As late fall and early winter snow accumulates, herds containing hundreds and sometimes a thousand or more animals congregate in wintering areas. This is the time of year when predators, especially coyotes and eagles, position themselves strategically to pick off the old, weak, and crippled. On two different occasions in late winter, I witnessed weakened pronghorn killed by pairs of golden eagles. Both times, one eagle flew over the animal to distract it while the second one lit on the animal's back and ripped open the hide. Then both began to consume the exposed flesh before the animal died. Other eagles soon joined the first two; within a few hours, little remained except hide, head, bones, feet, and a few scraps rapidly being eaten or carried away by ravens and magpies. The late William Maycock, a long-time hunter and outfitter in eastern Wyoming, observed an almost identical occurrence, and, having considerable artistic talent, he painted a picture of the event (figure 28). My wife, June, and I watched a pair of golden eagles employ the same cooperative strategy on a jackrabbit. Coyotes are also efficient predators; operating singly or in groups of two or more, they find old and weakened pronghorn especially easy prey.

Ironically, late winter, when the faint promise of spring returns, is the

FIGURE 29. Frozen Wyoming pronghorn
retrieved from a herd of eleven that wandered
into a snow-filled arroyo. (Photo by author.)

time when most wild animals die, falling to predators, malnutrition, and
bad weather (figure 29). The large winter herds of pronghorn begin to
disperse when the snow melts and the first spears of green grass appear.
On numerous occasions, I have watched both pronghorn and mule deer,
largely oblivious to my presence, literally exhausting themselves in their
pursuit of the earliest grass. However, as more green grass becomes available, they soon recover and show little resemblance to their shaggy late
winter appearance. The green grass of spring is unrivaled as a conditioner
to revive winter-weakened animals. As they rapidly gain weight, they shed
winter coats and acquire shiny new ones.

Males of all ages tend to form summer groups, as do the young and
barren females. Females carrying fawns become solitary in early spring
and select a secluded location where they will give birth. Both breeding
and birthing periods are relatively short, and fawns are usually born during late May and early June. Newborn fawns exude little odor, a natural protection from predators; and within a few hours of birth, a fawn
is a tiny body with long, thin legs capable of attaining surprising speed.
Females hide their newborns in brush or tall grass while they are away
feeding. When approached without their mothers close by, fawns usually stay motionless, with their heads pressed to the ground. However,
if prodded or otherwise forced to move, they will usually run away at
top speed. If they are very young and their mothers are not around to
claim them, they occasionally imprint on another animal such as a horse

or cow, or even a vehicle. In such cases, they can sometimes be difficult to separate from the object they become attached to.

Females produce ample quantities of milk, and fawns rapidly increase in size; soon small groups of females with young aggregate. As grasses mature in summer, animals acquire the fat reserves needed for the coming winter. Aggregates of male pronghorn, compatible through spring and early summer, now disperse as dominant males begin to show their authority with the approach of the rutting season. Mature males now collect females, usually five to ten in number, and seclude them in compact "harem" groups. They constantly face challenges by younger males; a common sight is a dominant male in very fast pursuit of a younger one for a distance of half a kilometer or more, but returning quickly to protect his harem from another male intruder.

Male pronghorn rutting behavior is perhaps best described by Maycock, whose education as a hunter began at about the same time as and is, in many ways, reminiscent of my own:

> Buck antelope lose much of their cunning while in the rut and try to keep their herds in a certain locality and much of their time is spent herding their charges to keep them from being spied by a rival. If a rival approaches, he is usually up where he can see him coming and will charge out to meet him before the interloper has found what a cozy nest he has. If a fight does not ensue, he will run this buck off and then saunter back as if nothing happened. Rubbing his horns on a sagebrush and pawing it as if to say the world is mine, who is next to challenge? (Maycock 1980: 85)

Pronghorn prefer open country where they can see long distances, and many migrate between summer and winter habitat areas that are widely separated. These seasonal migrations often require movements through areas of rough topography, forest cover, roads, streams, and livestock fences. A long-term study of pronghorn spring and fall migrations by the Wyoming Game and Fish Department in northwest Wyoming provides a good look at these spring and fall migrations. Radio collars were placed on 25 animals in a herd of about 100 pronghorn in order to track their movements between their summer locations in the sagebrush flats of Grand Teton National Park and their wintering area more than 160 kilometers away, along the Green River. The migration pattern varied slightly from year to year, apparently owing mostly to changes in annual weather conditions. For example, females were held up one year by an intense spring storm and forced to give birth to their young just short of their normal destination. In that case, and with adequate feed conditions, they spent the summer short of their normal summer range. However, when the weather was more favorable the following year, they negotiated the entire route. Their migration took them over a hydrographic divide more than 3,000 meters in elevation and through deep canyons and timbered areas (see Sawyer and Lindzey 2000).

I am personally familiar with a small herd of pronghorn in the Big Horn Basin of Wyoming that spends the winters at an elevation of 1,400 meters and the summers at about 3,000 meters in the Absaroka Mountains. It should be noted, however, that this pattern reflects the area's unusual ecology, in which open, grassy slopes provide unrestricted access from the interior basin to above the timberline. Depending on local conditions, other pronghorn populations may move short distances seasonally or occupy a single area year-round.

Pronghorn, deer, and mountain sheep are susceptible to an insect-borne, infectious, and noncontagious disease commonly known as bluetongue. The virus is carried mainly by the tiny biting gnat usually called the no-see-um *(Culicoides variipennis)*. There was a die-off of an estimated 4,100 pronghorn and deer in eastern Wyoming in 1976 (Thorne, Kingston, et al. 1982).

PRONGHORN HUNTING

Several behavioral traits largely determine the strategies used to hunt pronghorn in the present and, probably, in the past. Perhaps the most readily observed of these is their curiosity. My first experience hunting pronghorn was with an old rancher in the Powder River Basin of eastern Wyoming during the fall of 1940. I arrived with hunting gear, ready to strike off across country in pursuit of the animals, but he suggested that I relax and wait until the next morning. We arose well before first light and hiked to the top of a low hill a short distance from the ranch house; there he hung, on a stick, an old coat that could be seen for some distance in all directions as it waved in the breeze. At first good light, several pronghorn could be seen moving slowly toward us, stopping occasionally but always alert, watching the coat flapping in the wind. By sunup, there were several within easy shooting distance; and because he was more interested in meat than in a trophy, my companion pointed out a yearling male that was in good condition. One easy shot and the hunt was over for that year. Later on at home, the consensus was that pronghorn meat, at least from that particular animal, was not much different from deer.

The following year, I was determined to acquire an animal more in the trophy class and spent two days before hunting season familiarizing myself with an area reputed to be the home of a large male. I was unaware, however, that another hunter had the same intentions. Early in the morning the day the season opened, he spotted the animal first; but it was up and running immediately. He shot but missed, and the animal came by me at top speed. Possibly because I was used to hunting mule deer and did not allow for the pronghorn's much greater speed, my shot hit well behind him and he was quickly out of range. I spent two days searching the area and encountering numerous animals without sighting the big one again. I finally settled for a younger but still respectable trophy. The episode emphasized one facet of pronghorn behavior: when hunting pressure is on large males, they have an uncanny ability to find hiding places in patches of brush, shallow arroyos, and depressions in what appears to be almost featureless terrain. When flushed out of hiding, they rely on their superior speed to escape.

Although pronghorn blend well into the terrain and vegetation, they are especially visible just after sunup and before sunset, when the slanting rays of the sun reflect the distinctive coloring of their white rumps, which can be seen for long distances. When pursuing pronghorn or most other animals, the hunter has a definite advantage if he or she can locate the animals without first being seen by them. Stalking a pronghorn is usually a good strategy in rough terrain, tall grass, or brush cover, but frustrating in open, flat country where the animals can detect the movements of approaching predators far away.

Much of pronghorn habitat is in areas of low precipitation, little topographic relief, and limited vegetation, with scarce water sources for much of the year. Sitting tight in camouflaged hunting blinds near watering holes can be a successful hunting strategy—especially in hot, dry weather, when pronghorn travel back and forth to water daily and, under extreme conditions, more than once a day. For one hunter to hide in a blind while one or more others flush out animals and drive them by is a common strategy, especially effective for present-day bow hunters, because it increases the chances for relatively short-distance shots.

A behavioral trait of pronghorn that makes them entirely different from deer, elk, and mountain sheep is a stubbornness that manifests itself under certain conditions. For example, lone animals and small groups in open, flat country will often start running parallel to and in the same direction as a slow-moving vehicle; if the driver gradually increases speed, they will exert all possible effort to cross in front of it. Once this objective is accomplished, they usually run a short distance, stop, and look

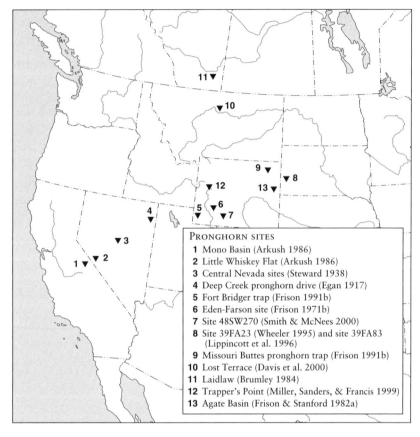

MAP 5. Selected pronghorn archaeological sites.

back at their pursuers. If the vehicle accelerates too fast during the process, the animals will veer away and abandon the attempt to outrun it. Pronghorn also show an out-of-sight, out-of-mind attitude toward hunters even when they are under intense hunting pressure. They usually run away from a perceived threat at top speed until well out of sight of their pursuer, and then apparently dismiss the danger and resume whatever they were doing when disturbed. In this situation, the best strategy for the hunter is to keep out of sight, allow the animals a short rest, and then resume the stalk. To elude predators, human and other, pronghorn rely strongly on speed, eyesight, and quick reflexes and less on scent than do bison and elk. Their curiosity extends to domestic animals; they are often observed grazing close to and unconcerned about cattle, sheep, and horses on the open range. Under severe winter conditions they are a common sight around the fringes of human habitation areas, as long as they are unmolested. When their feed is deeply buried under snow and ice, it is not unusual to look out a window and see pronghorns eating any exposed vegetation and pawing for food buried under the snow.

A field-dressed mature male pronghorn in good bodily condition with head, hide, and feet removed usually weighs about 36 kilograms. Females weigh about 7 kilograms less, and a healthy fawn in early fall should weigh between 16 and 18 kilograms. Their habitat is nearly always in areas of low topographic relief and sparse vegetation, so when they are killed their carcasses usually are easily accessible to the hunter. Even a large male pronghorn is not too great a load for the average hunter to shoulder and carry a long distance back to camp. Since pronghorn are fairly easy to locate, procure, and retrieve when killed, the success of single hunters and small groups over the long term is relatively high.

Pronghorn behavior makes it possible for humans to drive and otherwise manipulate them readily enough to trap them in relatively simple enclosures. The reluctance of pronghorn to go over or through simple barriers is the behavioral trait that made prehistoric communal human hunting easier. As is the case with bison, ethnographic and eyewitness accounts of past pronghorn trapping, along with remnants of trapping complexes utilized in the past and a close familiarity with their present-day behavioral characteristics, provide archaeologists with sufficient information to arrive at accurate reconstructions of past procurement strategies. Unlike prehistoric and historic bison kill sites, however, pronghorn sites of the same age that provide good evidence for procurement are limited (see map 5).

ETHNOLOGICAL AND HISTORICAL EVIDENCE OF PRONGHORN HUNTING

The Great Basin covers a large area of western North America and was occupied in prehistoric times by various Shoshonean groups. Eyewitness accounts give us a detailed picture of historic communal pronghorn procurement by Shoshoni in the region. The general plan involved traps consisting of strategically placed brush fences leading to circular enclosures that were of relatively simple construction; they were effective because the animals refused to jump or force their way through them. Once inside the enclosure, the animals could be run to exhaustion and killed with clubs.

Classic examples of this type of pronghorn trap are located in the

Mono Basin in east-central California (Arkush 1986: 248) and at the Little Whiskey Flat site in west-central Nevada (Arkush 1986: 247). According to Brooke Arkush, the latter site was first seen and described as a deer trap by a member of the 1845 Fremont expedition, but later archaeological investigations confirmed that nearly all faunal remains were pronghorn. Howard Egan (1917: 238–41) actually participated in a Gosiute pronghorn drive into a trap at Deep Creek in northeast Nevada in the late nineteenth century (see also Arkush 1986: 244; Reagan 1934). Julian Steward's accounts (1938) of Shoshonean pronghorn procurement are revealing in that he records the importance of the shaman's presence during communal trapping and the refusal to carry out a communal hunt if a shaman was unavailable. Steward also mentions communal hunts that could eliminate so many animals in an area that a period of years had to pass before their numbers recovered enough to allow another hunt. Communal hunts required that a minimum number of animals be available to justify the effort involved in constructing and maintaining drive lines and enclosures.

The acquisition of horses by the Wind River Shoshoni in historic times apparently changed their pronghorn hunting strategy. According to one eyewitness account, an estimated fifty riders were able to surround a pronghorn herd; the participants took turns running them in circles until they collapsed from fatigue and were then easily killed (E. Wilson and Driggs 1919). No shaman's presence is mentioned.

Robert Lowie presents a brief account of the Northern Shoshoni pursuing pronghorn using horses: one or two men would pursue a herd until their horses were winded and other hunters with fresh horses would resume the chase. This was repeated until the animals were exhausted, when they were killed with arrows. According to Lowie (1909: 185), "It would sometimes take forty or fifty hunters half a day to kill two or three antelopes by this method." He then quotes Lewis and Clark: "twenty men set out after a herd of ten head and were unable to capture a single animal in a two hour's run." If these are accurate accounts, I would have to interpret this tactic as more sport than serious hunting.

The extreme southwestern part of Wyoming is considered part of the Great Basin; the remains of a Shoshonean pronghorn trapping complex, known as the Fort Bridger Pronghorn Trap, are still visible there, in good pronghorn habitat, about 40 kilometers northwest of the small town of Fort Bridger (figure 30). It was constructed of juniper trees, but most of the larger logs were removed during the first part of the twentieth century by local homesteaders because they were an easily obtainable source

FIGURE 30. Location of the Fort Bridger, Wyoming, pronghorn corral. (From Frison 1978: 255.)

of firewood. Though what is left is rapidly rotting away, enough still remains to provide the details of the entire complex (map 6). The site is close to two playa lakes—a location where pronghorn congregate at present, especially after late summer and fall rains regenerate the vegetation around their edges. When a herd of desired size gathered, the animals were driven against the wing of the trap, which led them to the opening of the circular enclosure. Once inside, they were circled until exhausted and, given the lack of projectile points within the enclosure, were most likely killed with clubs. A juniper or sagebrush bark rope may have been added to the top of the fence to better ensure that the animals were contained. A small concentration of juniper tree limbs incorporated into the northwest part of the enclosure fence appears out of place and may have been part of a shaman's structure. Outside the enclosure is an exposed area that was almost certainly a camp or processing area; it has yielded Late Prehistoric–type projectile points, lithic tools, and a few badly deteriorated bone fragments, but no historic items.

The northwest side of the trap has a double fence (map 6) that most likely represents an effort at rebuilding to improve the function of the trap. The last stage of building was most likely accomplished largely by pirating brush from the earlier fence, but the more recent removal of most of the wooden parts for firewood has made it difficult to determine which was the abandoned section. That parts of a complex of this nature were rebuilt does not suggest a poor original understanding of pronghorn

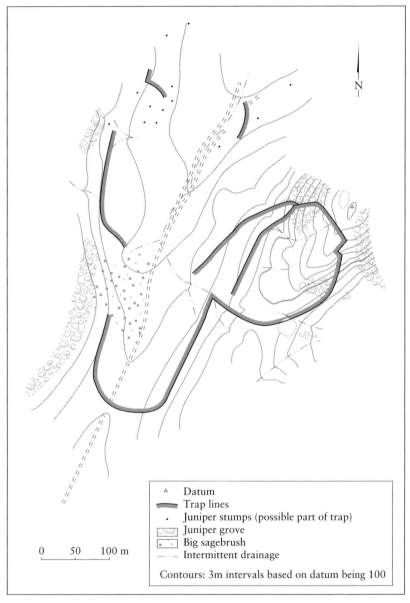

behavior. In my own experience, I have found that trial and error usually reveals unexpected problems in corrals built for domestic livestock on the open range; the same would apply to those built to contain wild animals. Relatively simple modifications usually rectify the situation; in rare cases, the original site has to be abandoned and a new location selected. There are no known eyewitnesses to the use of the trap, but judging from the deterioration of its wooden parts, its last use most likely dates to at least late historic times, toward the end of the nineteenth century.

In the late 1960s, an elderly resident of Fort Bridger, Wyoming, said he had seen this pronghorn trap and another one of similar size and shape in the same general area but constructed of sagebrush. He said he also remembered a man who claimed to have witnessed Shoshoni on foot attempting to drive pronghorn into the latter trap. If the account is reliable, this hunt would have occurred near the end of the nineteenth century. The informant provided a good description of the trap and its location, but his age and health prevented him from guiding me to the area. Although I searched the area thoroughly, I could locate no remains of it; my failure was not surprising, because sagebrush would have deteriorated much sooner than juniper. Intensive use of the area by domestic livestock over the past hundred years is another factor that very likely hastened the trap's disappearance.

The known pattern of historic pronghorn hunting on the Great Plains involved the use of horses. George Hyde (1974: 21) describes a Brule Sioux pronghorn trap in the Sand Hills area of Nebraska. In this account, hunters on horses drove the animals through a gap in the hills and over a low cliff into a brush-and-log enclosure; there hunters waiting outside the enclosure killed them with bows and arrows. The historic Cheyenne were observed using horses in communal pronghorn procurement, accompanied by shaman activity. Such accompaniment might be expected, because a communal pronghorn hunt had about the same possibility of failure as a communal bison hunt, making supernatural help welcome in both cases. Two drawings by Howling Wolf, a Southern Cheyenne, depict hunters on horseback with lances and with bows and arrows pursuing pronghorn (reproduced in Sundstrom 2000: 122–23). Driving pronghorn into excavated pits was apparently also part of the Cheyenne procurement strategy (Sundstrom 2000: 128).

Remains of juniper brush fences terminating in a pit are still present at a location in the Little Missouri headwaters area in northeastern Wyoming on the western edge of the Black Hills (Frison 1991b: 245). This trap complex, known locally as the Missouri Buttes pronghorn trap, was undoubtedly historic; out of the area of the Shoshoni, it was most likely Cheyenne, though it might have been of Sioux or Crow origin. A Cheyenne pronghorn pit was reportedly still visible in 1977 within the city limits of Belle Fourche, South Dakota (Stands in Timber and Liberty 1967: 84–85; Sundstrom 2000: 126). Sioux pronghorn procurement strategies would have developed over a relatively brief time, because their original home was to the east of the Missouri River beyond the range of pronghorn habitat (see Walker 2000); their approaches were most likely copied from tribes already residing west of the Missouri River.

PREHISTORIC PRONGHORN PROCUREMENT

Although in western North America pronghorn are commonly seen today and found in numerous archaeological contexts, relatively few pronghorn site studies have supplied the kind of data that enlighten archaeologists about actual prehistoric procurement strategies. Bison studies have overshadowed them, mainly because the skeletal remains of bison are better preserved and more visible, and more often provide adequate specimens for taphonomic analyses.

When the Union Pacific Railroad was built through southern Wyoming in the mid–nineteenth century, the railroad was given every other section of land 20 miles on each side of the railroad right-of-way. This checkerboard pattern of federal and private land covers a large part of southwest Wyoming that, past and present, is prime pronghorn habitat. The area contains reserves of oil, gas, coal, and other natural resources whose exploitation requires extensive surface damage. However, federal regulations governing cultural resource management require surveys and mitigation of harm to archaeological resources, and as a result many prehistoric archaeological sites in this area have been found that contain pronghorn remains. After information had been collected for more than three decades, a symposium was organized at Western Wyoming College in Rock Springs, Wyoming, to bring together and present as much data as possible concerning prehistoric human pronghorn procurement. The outcome was a good synthesis of the presently known data, which provides an excellent foundation for further research (see Pastor and Lubinski 2000).

The earliest investigation of a pronghorn site in the area was during 1969 at the Eden-Farson location about 50 kilometers north of Rock Springs at the western edge of the Killpecker sand dunes, one of the largest areas of active dunes in North America. Partial remains of more than

200 pronghorn were recovered in what is believed to be a Shoshonean campsite of one season's occupation (Frison 1971b, 2000b). Wind-blown sand drifted over the site and preserved the skeletal remains. Taphonomic analysis demonstrates that most of the animals were killed over a short period during the latter part of October and early November. It could represent a single use or multiple kill events (Nimmo 1971). At that time of year, pronghorn are still in good condition, beginning to aggregate in larger herds; it is thus an ideal time for communal hunting. There is no evidence of a trap in the vicinity, though the site is located in an area of tall sagebrush, well suited for constructing brush fences capable of containing pronghorn. A radiocarbon date of 1720 C.E. ± 100 years (RL-101) could place it somewhere in the seventeenth, eighteenth, or nineteenth century. As was true of the Fort Bridger pronghorn trap, no European items or evidence of horse remains or horse gear was in the cultural assemblage recovered at the site. And also like the Fort Bridger site, the Eden-Farson site has all the earmarks of pedestrian pronghorn communal procurement.

This site contains the only presently known skeletal evidence of a large Late Prehistoric– or Early Historic–age communal pronghorn kill in the southwestern Wyoming area. It is difficult to explain why similar sites have not been found, although poor bone preservation could be part of the reason. Another possibility is that the site represents an overkill from which the local pronghorn population was unable to recover enough to allow communal hunts in succeeding years. The pronghorn bone at the Eden-Farson site was excellently preserved because it is in a depression immediately to the leeward of a source of wind-blown sand. We can hardly conclude that Eden-Farson is the only communal kill site in the area: the Fort Bridger site described above is the kind of trapping complex that one would expect to yield the number of animals found at Eden-Farson, but apparently there the bones were exposed on the surface and disintegrated. Known pronghorn traps are in exposed locations that, unlike arroyo bison traps, allow only rare opportunities for skeletal material to be covered by sediments; this may well be the most important reason that pronghorn procurement sites have been lost. The data from other excavated sites with good integrity in southwestern Wyoming, such as site 48SW270, yield small numbers of pronghorn and strongly suggest "a less spectacular picture of pronghorn exploitation" involving "hunting individual pronghorn as part of a more generalized procurement strategy" (Smith and McNees 2000: 71; see also Fisher and Frison 2000).

There is a topographically determined bottleneck between pronghorn summer range as far north as Grand Teton National Park and crucial winter range on the open sagebrush country east of the Green River and north of Interstate 80 in western Wyoming. This constriction of the migration route lies on a ridgetop between the Green River on the west and the New Fork River on the east, at a location known as Trappers Point. Recent widening of a highway exposed a stratified archaeological site with Early Archaic components radiocarbon-dated to about 7,900 to 4,700 years ago, with pronghorn dominating the faunal remains. It is known as the Trappers Point site (Miller, Sanders, and Francis 1999), and one component dated at about 5,700 years ago suggests a catastrophic mass kill. An estimated one-quarter of the bone bed in this component produced a minimum number of twenty-seven individuals and, in addition, parts of eight fetuses. Tooth eruption and wear, along with the stages of fetus development, indicate that the animals were killed from the middle of March through April, a time period during which, judging from present migration patterns, the animals were beginning their move from winter to summer range. An alternative interpretation is that hunters, instead of waiting for migrating animals, drove pronghorn through this bottleneck well before the animals began their regular migration. This component strongly suggests a catastrophic kill; but whether it was a single event, more than one event at about the same time during one year, or multiple events at about the same time over more than one year cannot be determined on the basis of present evidence. In addition, it is not yet known if other site components represent pronghorn procurement at this early spring season.

Much of the Trappers Point site is still intact, although an unknown proportion has been lost to erosion and highway construction. However, the evidence strongly suggests long-term and repeated use of the location either to intercept seasonal migration patterns or to execute planned drives of animals resident to the area. Although one component suggests a communal kill involving large numbers of hunters and some kind of artificial trap, it is also possible that fewer hunters used simple blinds or topographical features of the location to ambush animals. One or more traps, similar to those known historically, may have been nearby; the lack of evidence is no surprise, given the several thousand years that have lapsed since the last known use of the site. However, this site provides strong evidence of hitherto unknown, but long suspected, time depths for both human procurement of pronghorn and patterns of pronghorn migration. The surface evidence of lithic assemblages suggests the likelihood that this pronghorn procurement strategy began even earlier, in Late Paleoindian times. This hypothesis might be proven by further investigations at Trappers Point if it contains as yet undiscovered older components.

Included among the faunal remains at the Trappers Point site, and presumably utilized for food, were small numbers of bison, elk, ground squirrel, cottontail, and sage grouse. With few exceptions, all skeletal elements demonstrate intensive breakage to salvage all edible parts, in contrast to many plains' bison kill sites where carcasses were used less intensively (see Todd 1987).

The Lost Terrace site along the Missouri River in northern Montana portrays a single winter's exploitation of pronghorn by Late Prehistoric–age Avonlea hunters, probably about 1,200 years ago. The investigators (Davis, Fisher, et al. 2000) suggest one or more communal winter kills in a herd of pronghorn forced to survive on a river terrace. As was also true of the Trappers Point pronghorn, seasonality was determined by tooth eruption in postnatal animals and by the development of fetal bones. It was not the kind of location where pronghorn normally choose to spend the winter: however, I have observed pronghorn in both Colorado and Wyoming forced into similar situations by unfavorable weather conditions. Blizzards cause them to drift into arroyos or up against tight fences. A few warm days followed by intense cold forms a thick crust of ice on the deep snow, causing the animals to exhaust the available feed and become too weak to paw for more or move to another area. Unless they gain relief from a spell of warm winter weather, often in the form of a wind commonly referred to as a *chinook,* their chance of survival is extremely low.

In the case of the pronghorn at the Lost Terrace site, an artificial trap may not have been necessary. The location of the site on a river terrace and the timing of their death suggest animals unable to gain access to their normal winter feeding grounds. It may have been a relatively simple matter to kill animals weakened by lack of food and cold weather. I have observed pronghorn in late winter, trapped by fences along highways, that are too weak to move more than a short distance; when disturbed, they lie down and never regain their feet. This or something similar may very well have been the scenario at Lost Terrace.

Perhaps the only positive result of present-day large winter kills of pronghorn is that they provide archaeologists with large samples of comparative material for taphonomic study. Data on population structures derived from a large herd of pronghorn dying throughout the winter and another from a large herd killed within the space of a few days are different enough to convince most investigators of the value of taphonomic analysis as a means of determining seasonality and the population structure of animals recovered in communal kills.

Late Prehistoric evidence of pronghorn procurement is known from two sites, 39FA23 and 38FA83, both in the Cheyenne River area at the southern end of South Dakota's Black Hills. The first investigator (Wheeler 1995) found evidence of at least twenty-one pronghorn at the 39FA23 site, and a later investigation (Lippincott and Byrne 1996) yielded fifteen more. The site is thought to be a result of Plains Village groups making late-summer or early-fall hunting forays into the southern Black Hills area (Lippincott, Adair, et al. 1996: 102). Provided the sample of pronghorn remains is adequate, taphonomic analysis could strengthen the seasonality determinations and also give clues as to the actual strategies employed to procure pronghorn. According to the investigators, only a small part of the site has been excavated, so the potential to obtain a large skeletal sample of pronghorn remains appears to be high. The other site, 39FA83, yielded remains of twenty-nine mature and nineteen immature pronghorn, but numbers of individual skeletal elements are not given (Wheeler 1995: 188). The quantity of bone leads one to expect that analysis could yield information on time of year and type of kill. These two sites strongly suggest intense Late Prehistoric pronghorn procurement activity in an area with large numbers of animals present then, as they are today. Pictograph Cave along the Yellowstone River near Billings, Montana (Mulloy 1958), produced small numbers of pronghorn bones throughout cultural deposits dating from Early Middle Prehistoric (Middle Archaic) times, estimated at about 5,000 years ago, to historic times (Olson 1958). This is an area of the plains that now supports large numbers of pronghorn.

The Laidlaw site, in southern Alberta near the town of Medicine Hat, is with little doubt a pronghorn trap dating to about 3,000 years ago (Brumley 1984). Lines of stones mark drive line fences that converge at a small catch pen that appears to have been a pit, although part may have been a wooden structure constructed above ground. Inspection of the site area convinces me that its location was determined by topography favorable for driving pronghorn. The site provides a time depth well into the Archaic period for communal pronghorn hunting on the plains.

Evidence of Paleoindian pronghorn procurement is limited. One ulna was recovered at the Colby Mammoth site (Walker and Frison 1980),

FIGURE 31. Mixed bison and pronghorn bones in a Folsom component at the Agate Basin site, eastern Wyoming. (From Frison 1982b: 41.)

and a cut-off proximal end of a metatarsal was recovered at the Shea-man Clovis site very near the Agate Basin site in eastern Wyoming (Frison and Craig 1982: 164). The Agate Basin site is located on the plains to the west of the southern end of the Black Hills. The Folsom component at the site contained partial remains of five pronghorn along with the remains of eight bison (figure 31). The pronghorn bones most likely represent animals killed elsewhere at different times throughout the winter and then brought to the site. Cutmarks, impact fractures, anvil damage, and burning indicate butchering and processing (Walker 1982; Hill, Frison, and Walker 1999). A small number of pronghorn bones also appeared at the Lindenmeier Folsom site in northern Colorado (Wilmsen and Roberts 1978). Both sites have radiocarbon dates of about 10,800 years ago on their Folsom components. The small number of known pronghorn bones in Paleoindian sites severely limits our information on procurement strategies.

Pronghorn have made a good recovery because of management practices that have allowed them to be transplanted into former habitats. Modern trapping methods are hardly reminiscent of the aboriginal ones, though they too rely strongly on animal behavior. While fixed-wing aircraft, helicopters, and high fences simplify these efforts, experience and extreme care are important to avoid unnecessary injury to animals. A good description of strategies used in modern pronghorn trapping is offered by Bert Popowski and Wilf Pyle (1982: 188–98).

In summary, pronghorn have been ubiquitous and little changed long before and throughout the more than 11,000 years of known human hunting of large mammals on the North American Plains and in the Great Basin. Their behavior patterns are different from those of other herbivores of similar size, such as deer and mountain sheep, requiring procurement strategies that took these differences into consideration. They were easy prey for hunters working alone and in small groups because they are amenable to being driven. They were taken communally by pedestrian hunters during Late Prehistoric times and by mounted hunters after the introduction of the horse. A growing body of evidence suggests communal hunting may have begun much earlier than previously suspected. Pronghorn were a reliable food source that, in some cases and in some areas, may have rivaled bison in importance.

CHAPTER 6

The Rocky Mountain Sheep

THE PRECARIOUS EXISTENCE OF A THREATENED SPECIES

Beginning at an early age and as the result of a multitude of unrelated events over several decades, I developed a lifelong fascination with mountain sheep *(Ovis canadensis)*. In the immediate area of the Big Horn Mountains in northern Wyoming where I was raised, they had been nearly eliminated by hunters soon after the beginning of the twentieth century. However, sheep horns and skulls were much in evidence, attesting to their former presence in considerable numbers, and a few managed to survive in the more inaccessible and rarely visited backcountry. A large ram silhouetted against the sky, or traversing what from a distance appeared to be a seemingly perpendicular cliff face, was a rare and thrilling sight. When I was about six years old and riding for cattle with my grandfather, he pointed to a location near the bottom of a steep canyon where he and another man had killed two rams in 1901. They salvaged the meat from the two animals and threw the skulls into a rock crevice, fully intending to retrieve them later. They were never able to revisit the location and the skulls remained there until a hiker found them, still well preserved, in the early 1960s. When I told him the story of their origin, he gave me one that I donated to the University of Wyoming's biological collections.

The area gained some notoriety for its sheep hunting in the late nineteenth century when a titled Englishman, Lord Gilbert Leigh, fell over a steep cliff and to his death in the canyon more than 100 meters below while in pursuit of mountain sheep. A crude stone monument on a high promontory overlooking the location that was visible from the old family ranch commemorates the event. For a young and impressionable boy, it added an element of intrigue and danger to mountain sheep hunting. It also whetted my desire to explore the canyon country, where I found numerous skulls, horns, and caricatures of mountain sheep, along with those of other animals, painted on and pecked into the walls of caves and rock shelters by earlier inhabitants.

One day when I was about seven and again riding for cattle with my grandfather, we stopped at a neighboring ranch for lunch and I noticed a strange looking animal in the horse corral. I soon learned it was a young mountain sheep that the rancher had found as a young lamb, somehow separated from its mother. He took it home, where it thrived on cow's milk and became a pet. Years later, the rancher related an interesting account of his further experiences with the young male mountain sheep. He kept it in the horse corral, which was made of pine poles about 20 centimeters in diameter, with ends stacked alternately one on top of the other to a height of about 2.5 meters. The corral was held together by paired posts set deep in the ground opposite each other, and separated from each other by the diameter of the poles. The young ram was soon able to put its feet through the openings between poles, climb to the top, and use the uppermost pole as a walkway. It was interesting and even entertaining to watch, until the animal learned that it could jump from the top of the corral to the roof of the barn adjacent to the corral, where, not content with simply gaining access, it used the barn roof as a place to run, thereby destroying shingles. To stop this access to the roof the rancher nailed boards on the sides of the poles to form a solid fence and prevent the animal from getting its feet between the poles and climbing to the top. This solved the problem, but only temporarily.

By this time, the rancher had developed a strong affection for the young ram but the relationship was strained because the animal was rapidly becoming a nuisance. It became a serious contest between man and beast over the barn roof when the ram discovered it could start at one side of the corral, take a run toward the board fence, jump as high as it could, and, when its feet came into contact with the boards, use all four feet to propel itself far enough upward to get its front feet over the top. It then once more could gain the top of the fence, and jump from there to the barn roof. As his final solution, the rancher dug along one side of the posts holding the corral upright and leaned the fence inward several degrees. Further attempts by the ram to reach the top of the fence were unsuccessful: it would run toward the fence, jump as high as possible, and hit it with its feet as before—but could not get any closer to the top and would only fall back into the corral. A short time later, the ram wandered away and was not seen again. Though this account may at first seem unconnected to mountain sheep hunting, it reveals a facet of mountain sheep behavior that later on helped me understand the operation of widespread Late Prehistoric and Historic sheep traps found throughout the mountains.

On another occasion, four cowboys riding in from the fall roundup in the mountains stopped by our ranch at supper time and told their own story about mountain sheep. While riding back to camp at dusk the evening before, they were surprised to come upon several sheep grazing in an open area. With two riders in the rear and one on each side, they drove the sheep into a pole corral a short distance away at their cow camp. However, before they could close the corral gate, the sheep had climbed over the rear of the corral and escaped, providing another bit of evidence of the ability of mountain sheep to easily negotiate barriers that other animals—including deer, pronghorn, and elk—cannot.

These experiences provided me intriguing information on mountain sheep but no opportunities for close contact with the animals. It was only after graduate school, when I was over the age of forty, that the situation changed, owing mainly to circumstances that brought me into contact with three individuals who all had long-term experience with, but different interests pertaining to, mountain sheep. One was Nedward Frost, a well-known mountain sheep outfitter and guide; another was Kay Bowles, an experienced game warden and wildlife manager; and the third was Roy Coleman, an outfitter and guide who took me to see wooden structures in the wilderness area of the Absaroka Mountains east of Yellowstone National Park.

Much of the Yellowstone Plateau and the surrounding mountains—now generally referred to as the Greater Yellowstone Ecosystem in Wyoming, Montana, and Idaho—is relatively inaccessible and has harbored mountain sheep herds that have been hunted for well over a century. The late Nedward Frost of Cody, Wyoming, was one of the more notable and successful mountain sheep hunting guides during the early part of the twentieth century. He was also well educated, a keen observer, a historian, and a writer. He catered to the trophy hunters, wasting little time on hunters who would be satisfied with a ram with a small set of horns. It was a big territory, and there were unwritten agreements among the major outfitters and guides to respect the boundaries between hunting areas claimed by each. All these guides resented intrusions by other hunters, and Frost was no exception. A common remark at the time was that if he discovered a horse track within two miles of his camp, he felt the area was becoming too crowded and he would immediately break his own camp and move further back into the mountains.

Although Frost's livelihood depended on sending hunters home with trophy mountain sheep heads, he had a deep feeling for the animals; and as hunting pressure continually increased, he became concerned about their survival. In 1940 there were only an estimated 3,000 sheep left in Wyoming, nearly all in the high, inaccessible mountain areas of the northwestern quarter of the state; and only a small number were in the Big Horn Mountains. The Wyoming Game and Fish Department decided to develop a long-range management plan, not only to prevent further loss but also to start the animals back on the road to recovery. Everyone who understood the problems confronting mountain sheep at that time also knew that Frost was more familiar than anyone else with the different mountain sheep herds still in Wyoming, and his expertise was wisely sought (see Honess and Frost 1942). His recommendations were taken seriously, with positive results; and, as we will see, his intimate knowledge of mountain sheep behavior was another element that helped me understand the procurement strategies likely utilized in prehistoric and historic communal sheep-hunting complexes.

The Absaroka and Wind River mountains in northwestern Wyoming, part of the Greater Yellowstone Ecosystem, are prime mountain sheep territory. Sheep have adapted to survival in inaccessible rough country, but their survival is constantly threatened. Wild sheep are highly susceptible to the lungworm parasite introduced by domestic sheep. Large die-offs of mountain sheep herds have been caused by pneumonia, a condition brought on by stress resulting from undernourishment, competition from domestic and other wild game animals, predators, extreme winter weather, and harassment by humans. Healthy sheep are better able to cope with parasites and disease, and it is a continual management problem to provide them with a healthy environment.

One mountain sheep management area known as Whiskey Mountain, located within the Dubois sheep trap area (see map 7, below), contains a herd of more than a thousand animals and produces enough increase annually to allow limited net trapping of surplus animals, which are transplanted into areas where herds were lost earlier (see Thorne, Butler, et al. 1979). By participating in some of these trapping events I was able to gain valuable insights into mountain sheep behavior and form a better basis to interpret strategies used in prehistoric mountain sheep procurement. Kay Bowles of Dubois, Wyoming, a dedicated wildlife manager and game warden whose main job was to oversee and maintain the welfare of the Whiskey Mountain sheep herd, introduced me to mountain sheep behavior in the wild. He knew the animals and the territory, probably as well as anyone; he was born and raised in the mountains and spent most of his time there either afoot or on horseback.

HISTORIC LOG STRUCTURES
IN THE HIGH COUNTRY OF NORTHWEST WYOMING

The late Roy Coleman of Cody, Wyoming, was another successful mountain sheep, grizzly bear, and elk guide and outfitter with an interest in the archaeology and the history of the greater Yellowstone Plateau area. His passion was to locate the trail that the Nez Percé Indians, led by Chief Joseph in 1877, took from the time they entered Yellowstone National Park, finally emerged from the Sunlight Basin, and then headed north toward Montana in an attempt to avoid the U.S. Army and seek refuge in Canada (see Beal 1963). In pursuing this effort over many years, he gained an intimate knowledge of some very inaccessible parts of the Absaroka Mountains in northwestern Wyoming. Coleman maintained a hunting camp in the Sunlight Basin, located east of Yellowstone National Park and surrounded by prime hunting country. Learning of my interest in the archaeology of the high country, he provided horses and mules for a look at caves and rock shelters just below the timberline. He also led me to a cribbed log structure 2.7 meters by 6.7 meters at its bottom with sides and ends converging as they rose upward. It stands 1.8 meters high but was originally higher (the bottom logs are now almost entirely rotted away). It is situated just inside a patch of thick lodgepole pine timber, close to an open grass-covered flat. Its size, shape, and lack of cultural materials, inside or out, eliminated it from consideration as a structure for human occupancy, but there was nothing associated with it to provide a clue to its former use.

I was convinced by this time that a wealth of archaeology was to be found in the high country (figure 32) that constitutes much of the Greater Yellowstone Ecosystem in northwestern Wyoming, southwestern Montana, and southeastern Idaho. Kay Bowles informed me of wooden structures he had seen in the mountains that he suspected were once used to trap mountain sheep. Following his directions, I backpacked into the area

FIGURE 32. Part of the area of mountain sheep traps in northwest Wyoming.
(From Frison 1991b: 250.)

and located two of these that were exactly as he had described. On viewing two structures (figure 33) similar to the one seen earlier but with associated wooden fences and drive lines, I immediately realized that these and the Sunlight Basin structure shown to me by Coleman were the remains of animal traps. At that time I lacked enough familiarity with mountain sheep to come up with more than superficial explanations of their operation. This changed over the next few years as I came to be closely associated with Kay Bowles, Roy Coleman, and Ned Frost.

After Frost retired from the guiding business, he was hired by the state of Wyoming in 1967 as a historian to evaluate historic and archaeological sites for possible inclusion in the newly conceived National Register of Historic Places. Working closely with him as Wyoming state archaeologist, I expressed my interest in the animal traps located in the high country. Although no longer as agile as in his younger years, he immediately offered to acquaint me with some of the mountain sheep areas he knew and to look into the feasibility of the structures that I had visited earlier being parts of mountain sheep trapping complexes. He mentioned having seen some of these earlier, but he had not been able to study them closely because at the time he had been guiding hunters who were interested only in acquiring large animal heads. By this time I had lo-

FIGURE 33. Wooden remains of the Bull Elk Pass sheep trap, northwest
Wyoming. (From Frison 1978: 268.)

cated nearly a dozen of these wooden structures and had been told of
several others. One site in the southern part of the Absaroka Mountains
was of special interest: it was located on an isolated feature known as
Black Mountain, which rises more than a hundred meters above the surrounding area and has steep, rocky, forested slopes. A relatively flat area
of several acres on top of the mountain contained the remains of two
animal traps (figure 34) and the partial remains of a third. Still another
trap is located nearby, at a strategic spot on a steep slope of a deep canyon
(figure 35).

The traps were originally constructed of dead lodgepole pine, fir, and
juniper timber and showed no evidence of metal ax marks. Logs with
roots intact indicate the use of fallen trees, plentiful nearby at present
and probably also in the past. Some parts of the traps were more deteriorated than others, but enough remained to provide most of the details
of their construction. With some difficulty, Frost made the steep climb
to the top of the mountain, looked at the traps, and made a careful survey of the surrounding terrain. After about an hour of intense concentration, he pointed to a bare ridge about 300 meters to the west of the
two traps and bet me a steak dinner it was the location of a mountain
sheep bed ground. Because his prediction proved correct, he collected on

FIGURE 34. The Black Mountain sheep trap, northwest Wyoming. (From Frison 1978: 266.)

FIGURE 35. Mountain sheep trap located on the Wiggins Fork of the Wind River in northwest Wyoming. (Photo by Sally Wulbrecht.)

that dinner. The ridge was covered with sheep feces from long-term use; it is still used when sheep move there during the fall after spending the summer at higher elevations.

After careful inspection of the sheep bed ground and the surrounding terrain, he said that hunters could keep out of sight until they were in the proper position before making their presence known to the animals. He pointed out that upon leaving the bed ground, the sheep would immediately run downslope a short distance, make a half circle, and then head back uphill. He next suggested, predicting almost exactly where it would begin, that I follow one of the drive line fences that was strategically placed to intercept the sheep as they completed their half circle. From that time on, Frost was hooked on sheep traps and became a constant source of information on mountain sheep behavior. Unfortunately, he died a few years later, before we were able to visit more than a few of the trapping locations he had encountered during his hunting and guiding career.

ROCKY MOUNTAIN SHEEP BEHAVIOR

The present-day net trapping strategy in the Whiskey Mountain Management Area is successful largely because of the discovery that mountain sheep have an insatiable desire for apple pulp and cured bright green alfalfa hay. Late in the winter, some of the hungrier and less wary ewes will approach to within a few meters of a person distributing the bait. As large drop nets are suspended several meters off the ground, the sheep are decoyed into position by the bait scattered beneath them; a net is released when the desired numbers of animals are directly underneath it.

The reactions of sheep restrained by the net reveals interesting and important facets of their behavior. Most ewes and lambs struggle violently for a short while but then become relatively passive and wait for whatever will happen next (figure 36). Some measure of restraint is used to prevent animals weakened to different degrees by winter range conditions from becoming exhausted, but in doing so care must be taken to avoid injuring them. Ribs are easily snapped and muscles damaged; particularly vulnerable are the sheep's eyes, which protrude more than those of most animals. What appears to be a superficial injury can often result in the animal going a short distance when free of the net, lying down, and never regaining its feet. This kind of reaction, along with numerous others, resembles in many ways that of their Old World relatives, the introduced domestic sheep.

Coyotes and mountain lions are the predators most dangerous to

FIGURE 36. Mountain sheep trapped under a drop net. (From Frison 1987: 206.)

mountain sheep. When threatened by coyotes, sheep retreat into "escape cover," usually rough terrain or timber where their pursuers are at a disadvantage (see Thorne, Butler, et al. 1979: 51–52). In late August of 1982, while backpacking in the Absaroka wilderness area, I happened at sunup on a group of ten ewes and lambs on an open flat just above timberline. While observing them through field glasses, I saw a female coyote and her four pups start running toward the sheep, who immediately sought the protection of a steep rocky outcrop. Once there, the sheep stopped; the coyotes gave up the chase and, apparently realizing the futility of further pursuit, moved on to seek more vulnerable game.

Mountain lions have increased in number in the past three decades, and their superior ability as predators is becoming a serious problem for sheep populations. There are occasional reports of golden eagles carrying off newborn lambs, but wildlife biologists believe that their overall threat has been exaggerated. As I write this, it is too soon to evaluate the impact that the recent introduction of wolves will have on mountain sheep populations in the Greater Yellowstone Ecosystem.

MOUNTAIN SHEEP HUNTING

Without an in-depth familiarity with mountain sheep behavior, my first approach to sheep hunting was to take several days' food and start hiking through the high country until a legal animal was encountered. Only mature males can be legally killed, and applications far outnumber the permits allowed. Large males have therefore become difficult to locate when the hunting season begins, while ewes, lambs, and immature males are nearly always in evidence. For more than fifteen years, I applied annually for a permit; when I finally drew one, I spent eleven difficult days combing the high country before finding and procuring a legal ram. I never applied for another permit because I thought it far more rewarding to observe the animals, learn their habits, and apply the information gained to the interpretation of prehistoric sheep traps. I donated the skeleton of the one ram I killed to the University of Wyoming comparative osteological collections, where it serves a continuing and useful purpose. One thing I learned from this hunting experience confirmed what many sheep hunters had claimed: the meat is of excellent quality and taste but without the strong flavor associated with domesticated sheep.

After this hunt, I discussed it with Frost and asked him the secrets of his success with trophy sheep hunters. He answered that there are about the same number of strategies as there are individual hunters, depending on a host of conditions that include weather, time of year, vegetation, topography, and the animal's state. His most surprising revelation was that because his clientele sought only the largest rams, his favorite and most dependable strategy was to spend time in summer locating old rams that usually stayed by themselves and displayed somewhat unusual patterns of behavior. These animals usually come out early and graze until midmorning, then go into a thick patch of timber or the shelter of a big rock and sleep until midafternoon, when they again come out to graze until dark. He claimed his most reliable strategy was to watch a ram with glasses and spotting scope to determine the exact location where he was sleeping during the day. Knowing this, he and his hunter would carefully slip up on the sleeping animal and often kill it at a distance of less than 20 meters. It doesn't sound very sporting, but for him it was an effective approach for taking trophy rams.

The sheep hunter soon learns that repeated success in hunting legal rams requires spending long hours with glasses, locating them before they locate him. Mountain sheep have excellent eyesight, and when a big ram spots a hunter first—which they usually do, unless the hunter is extremely cautious—the ram usually disappears into rough country or timber, making pursuit of that particular animal futile until another day. But in addition, and confirming Frost's observations, nearly every sheep hunter I

FIGURE 37. Mountain sheep rams grazing in early May. (Photo by author.)

have interviewed has a special strategy for success under particular conditions imposed by weather or terrain.

Mountain sheep hunting in Wyoming is closely regulated to ensure the viability of herds. Except in rare cases, legally hunted animals are mature rams, usually seven to eight years old. A hunter has to decide quickly, and from a distance, whether or not the animal is legal to kill; the criterion used is the extent of the curl of the horns. Usually a mature ram will have a three-fourths curl; in other words, a straight line from the base of the horn to the tip passes through the forward part of the eye. Obviously, the hunter must be able to view the head from the side to make a proper determination and not kill an immature and thus illegal animal.

As a result of decades of only mature rams being hunted, the animals have become extremely wary during the hunting season. They abandon much of this wariness as the rutting season begins; for that reason, hunting seasons end just before the rut. As the first spears of green grass appear in the spring, rams lose much of their fear of humans and are particularly vulnerable. For example, I came across eight rams within 20 meters of a well-traveled highway on May 1, 1989 (figure 37). Two were well into the legal age range; during the fall hunting season, these animals would be eagerly pursued by hunters and difficult to locate. I stopped along the highway and walked to within 5 meters of them before they moved away to graze on another patch of grass.

ARCHAEOLOGICAL EVIDENCE
OF PREHISTORIC AND HISTORIC SHEEP HUNTING

A close inspection of the mountain sheep trapping area (figure 32) at different seasons rapidly convinces one of its appeal to mountain sheep and other large mammals. It is an ecologically diverse area with elevations from 2,100 meters to more than 4,000 meters. There are deep canyons, flowing streams, perpendicular bluffs, steep slopes, and an occasional area of moderate topographic relief; areas of thick timber adjoin scattered trees and shrubs, and these may change rapidly to areas with a thick cover of sagebrush and still others of open productive grassland. Rapid changes in elevation enable animals and humans to move quickly from extreme weather conditions at high altitudes to the protected areas below. Within a short radius of Dubois, Wyoming, and in the heart of this area, I have recorded eleven mountain sheep traps (see Frison, Reher, and Walker 1990) and have been informed of others.

There are a limited number of archaeological sites that reveal much of sheep hunting (map 7). After several years of closely observing present-day mountain sheep in their natural habitat and after many discussions with individuals long experienced in hunting and in guiding sheep hunters, I was convinced that the trapping complexes described above were designed for the express purpose of communal procurement of mountain sheep. Because there were still wooden components present, their age was obviously not very great, but it was still uncertain. Among the known sheep traps, the best approach for making accurate age determinations appeared to be tree ring dating. At the Bull Elk Pass trap, the most complete and best preserved of the known structures (figure 33), the entire circumference of a pole, a part of the original corral that was placed in the fork of a living pine tree, was completely enclosed by the pine tree's subsequent growth before it died. Fortunately, the tree ring sequence of the dead pine tree could be correlated with that of a living tree nearby; and because the presence of the pole affected the tree rings, the number of years between the time the pole was placed in the tree fork and the time the living tree was tested could be calculated to within about two or three years. Assuming that the pole was placed in the tree fork when the trap was constructed, the date of that construction was in the last decade of the eighteenth century.

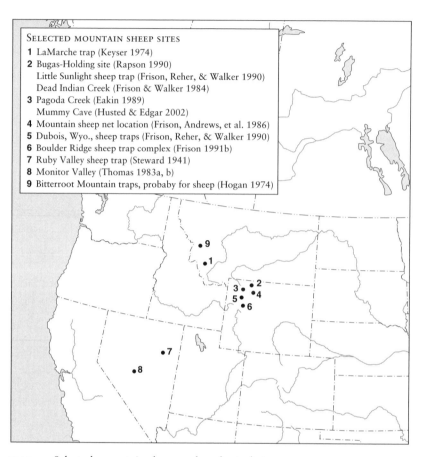

MAP 7. Selected mountain sheep archaeological sites.

Charles Reher's calculations of the number of rings from a living tree on the inside of a catch pen put one of the Black Mountain traps mentioned above (figure 34) at about the same date (Frison, Reher, and Walker 1990). Some other complexes in the area appear to be at about the same stage of deterioration as the dated ones, strongly suggesting that all were constructed in about the same time period. None of these traps is well enough preserved for use in its present condition: some wooden components above ground have collapsed, and any in contact with the ground have decayed to varying degrees. It is at this point that observations of the present-day behavior of sheep when corralled can provide valuable insight into the procurement strategy involved.

The effort needed to move, even over short distances, timbers the size of those used in constructing the traps soon convinces one that they were not constructed for occasional procurement of small numbers of animals. The density of trapping complexes in the study area appears to be higher than could be justified by the number of sheep there now. However, there is a body of data indicating that large numbers of mountain sheep were present at or close to the same time that the traps were in use. The Robert Stuart expedition of 1812 came up the Snake River in eastern Idaho, crossed to the Green River in western Wyoming, went on to the Sweetwater River, and followed the latter to the North Platte River. Part of this area includes the Ferris, Seminoe, and Casper mountain ranges in central Wyoming, within which Stuart (1935: 188–91) observed large numbers of sheep.

In 1887, John Fremont (1887: 133) in his journal notes large numbers of sheep near the confluence of the North Platte and Sweetwater rivers. In the same journal (1887: 147) large numbers of sheep north and west of that spot, in the Wind River Mountains, are also mentioned. Likewise, Osborne Russell (1921), a trapper traveling in the Wind River country of Wyoming in the summer of 1835, kept a detailed journal in which he made numerous references to large numbers of sheep throughout the area; he also reported on the ease of killing them.

Among settlers, J. D. Woodruff was one of the earliest in the southwestern Big Horn Basin of Wyoming in the mid–nineteenth century. In visits with my grandfather many years later, the subject of mountain sheep came up regularly. Woodruff said that during the last half of the nineteenth century, sheep were so plentiful that any time they were short of meat they hitched up a wagon, drove along the base of the steep east slope of the Absaroka Mountains, and loaded the wagon with sheep as they were shot and rolled to the bottom.

These accounts provide evidence that mountain sheep numbers may well have been sufficient in Late Prehistoric and Early Historic times to justify the construction of trapping complexes. Close contact with living mountain sheep convinces me, moreover, that the operation of the traps I have seen is well suited to their behavior. There were two basic types of traps: in one, the animals were apparently driven between converging drive lines directly into catch pens; in the other, they were first driven into holding pens and then into catch pens. The most likely advantage of the latter is that several animals could have been driven into a holding pen, with some of them then split off and driven into the catch pen; in this way, the animals could be dispatched a few at a time. It is possible that the second, more complex type of trap developed out of the first.

Each of the complexes investigated shared several characteristics indicating that the hunters were well versed in the behavior of mountain sheep. For the first type of trap, animals were driven uphill between converging drive lines until they were within a short distance of the catch pen. For the final push, the animals were driven downhill between drive line fences that terminated at the base of a ramp at the catch pen's entrance. In the other type of trap, the animals were driven uphill into the holding pen and then downhill into the catch pen. The catch pens were constructed of whole logs, some up to 30 centimeters in diameter and with parts of the root system attached, which added significantly to their bulk but also aided in holding the structure together. In some cases, catch pens were anchored to a rooted tree if one was properly situated. In the absence of such a tree, the pen was braced with large timbers. Catch pens were wider at the bottom than at the top to prevent the animals from climbing out. A ramp was built at the entrance to the catch pens, and it would appear to the animals that their only escape was to jump off the end of the ramp into the catch pen. The ramp was constructed so that they could not see the catch pen or any obstruction that might encourage them to turn around and attempt to flee. The floor of the ramp was covered with dirt and rocks, presumably to make it appear as natural rather than artificial terrain.

Instead of building a ramp, the builders of the catch pen at the Bull Elk Pass trap took advantage of a large rock at the site (figure 33). The animals were forced to jump more than a meter to the top of the rock and from there into the catch pen. In this instance, the catch pen and drive lines into the trap appear to have been designed to accommodate the rock: not only was it in an advantageous position, but it was far too large to be moved.

A well-preserved section of fence in one complex consisted of horizontal logs rising about a meter above present ground level and held in place with wooden uprights in shallow holes. A solid wall of poles placed upright against the horizontal logs made it lean toward the inside of the pen, offering more evidence that the hunters were well aware of a mountain sheep's ability to climb up and over a perpendicular fence and bringing to my mind the old rancher's account, told to me several decades earlier, of leaning a fence inward to keep a pet ram from climbing over the top. Remains of small wood and stone structures incorporated into drive lines that do not appear to affect the traps' functional utility are believed to have been associated with a shaman, who would have used them in ways similar to those observed during historic communal pronghorn procurement events.

FIGURE 38. Tree growth partially enveloping
a large ram skull. (From Frison, Reher, & Walker
1990: 233.)

Other evidence of shamanistic activity associated with the sheep traps
consists of large, mature ram skulls placed in the forks of trees. In some
cases, subsequent tree growth enveloped parts of the skulls (figure 38); in
other cases, the trees were dead at the time. Most of the skulls show that
the brain case was opened by smashing the foramen magnum, a further
suggestion of ritual activity. While traveling along the Bitterroot River in
the Flathead Indian country of southwest Montana with the American
Fur Company, Warren Angus Ferris noted a ram skull partially embedded in a tree. He mentioned that the local Flathead Indians left offerings
when passing by the tree (Ferris 1940). George Weisel, another observer,
mentioned "medicine trees" and cited an account by Alexander Ross, a
fur trader, who in 1824 may have seen the same tree Ferris described;
Ross also noted that the Indians always left offerings there when passing by. In addition, Weisel observed sections of two other trees with embedded mountain sheep skulls; beads and other trinkets were revealed
at the base of one of them when it was cut about 1920 (Weisel 1951:11).
Ritualistic treatment of large ram skulls was apparently a widespread

phenomenon among Native American sheep hunters in this part of the Rocky Mountains.

Skeletal remains of sheep in and around the known sheep traps are scarce, leading me to believe that once the animals were killed they were carried away to be butchered and processed at nearby campsites. Supporting this idea are two large Absaroka Mountain winter campsites, the Bugas-Holding site of Late Prehistoric age and the Pagoda Creek site dated to the Late Archaic (see map 7). The former is located along Sunlight Creek, a stream in the lower elevations of the Sunlight Basin in northwestern Wyoming, which produced mountain sheep remains killed in early fall (Rapson 1990). The sheep trap first shown to me by Roy Coleman is nearby, but at a much higher elevation and placed at a location favorable for trapping animals. It could have been a trap used by the hunters camped at the Bugas-Holding site. I believe that these sheep trapping complexes were designed mainly for taking nursery herds rather than rams.

The Late Archaic–age Pagoda Creek site is near the Bugas-Holding site as the crow flies, but access on foot requires a long trek over extremely difficult terrain. It lies on a terrace along the North Fork of the Shoshone River canyon. It is a Pelican Lake cultural complex site, dated to around 3000 B.P., and contains mostly mountain sheep remains killed within a short period in early winter (Eakin 1989). There is a strong possibility that the mountain sheep were being taken nearby in communal efforts, though evidence of traps has not been found. The faunal remains represent animals taken from nursery herds. Much of the site remains intact and has the potential to reveal significant information on mountain sheep procurement in the Late Archaic age.

In 1988, an elk hunter reported a large ram skull in the fork of a dead pine tree located deep in the Absaroka Mountains in northwestern Wyoming. I visited the location two years later; in the meantime the tree had fallen, but the ram skull was recovered intact. Time constraints permitted only a superficial investigation, which revealed badly deteriorated remains of log fences and at least twenty-five ram skulls along with some postcranial remains, also in badly deteriorated condition. This site is unique; because there is a strong indication of buried deposits, further excavation may help explain its presence. The ram skull in the tree suggests some associated ritual activity.

Live body weights of sheep trapped from the Whiskey Mountain herd are as follows: the average of sixty-nine ewes at least three years old over a three-year period was 124.3 pounds (56.4 kg) and for six males was

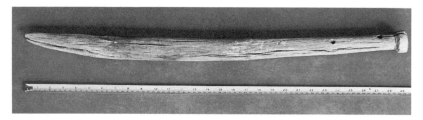

FIGURE 39. Wooden club found in a mountain sheep trap. (From Frison, Reher, & Walker 1990: 231.)

162.5 pounds (73.7 kg). Eighteen female lambs averaged 58.9 pounds (26.7 kg) and twenty male lambs averaged 60.5 pounds (27.4 kg), all at about eight months of age (Thorne et al. 1979: 74–77). Weights would be somewhat less for field-dressed animals; although the mountain sheep are somewhat larger than pronghorn, a dead animal could easily have been transported to nearby camps by a single hunter.

A noticeable lack of projectile points in the catch pens of the sheep traps strongly suggests that clubs were used to kill the animals. Shooting arrows at highly agitated animals in the confines of an enclosure would be extremely dangerous to the other hunters, and clubbing the animals to death seems a better strategy. A wooden club reportedly found inside the catch pen at one trap site resembles a modern softball bat (figure 39). Although I was unable to locate two other potential clubs claimed to have been recovered at another site, I was told that one was a stick similar in size to the one described above, wrapped with rawhide its full length; the other was a stick with an enlarged pathological growth on the distal end, the handle being wrapped with rawhide. Another probable club found near a sheep trap was a 70-centimeter-long proximal end of the main beam from a large elk antler with a sharpened bez tine. The wooden club pictured and the one made of elk antler would be entirely adequate for dispatching mountain sheep confined in the catch pens.

In addition to these large trapping complexes, the remains of isolated small rock and wooden structures located at strategic spots near water sources, along trails, and on promontories, as well as in and at both ends of mountain passes, attest to past hunting efforts by individuals and small groups of hunters. However, not all of these necessarily are connected with mountain sheep hunting—mule deer and elk also frequent most of these areas. But I do not believe the operation of the communal traps described above fits into the behavioral patterns of deer or elk.

This regional pattern of mountain sheep procurement is often attributed to Shoshonean groups commonly referred to as the Sheep Eaters
(see Dominick 1964), though the relation of the Sheep Eaters to other
Shoshoni is controversial (see Hughes 2000). A persistent theme among
people I have known personally on the Wind River Reservation in northern Wyoming, some claiming actual Sheep Eater ancestry, is that "Sheep
Eaters" is a derogatory term applied by the Wind River Shoshoni to other
Shoshoni confined to the mountains. The former acquired horses, traveled widely, and hunted mainly bison, while the latter had no horses and
hunted mountain sheep. The name may sound derogatory, but in my opinion more skill is required to procure mountain sheep than bison. Great
Basin Shoshoni may have used traps reminiscent of those described above.
In describing the Ruby Valley area of Nevada, Julian Steward (1941: 329)
mentioned a circular enclosure, 33 meters in diameter, "of mountain mahogany sticks sloping inward." It is also "near spring and has wings."
Steward added that "Enclosures, traps, nets, and snares were believed
to be ineffective." He also claimed, "Ambushing, stalking or driving onto
cliffs were most effective. For the last, dogs, which were little used for
other hunting, were employed" (Steward 1941: 220). I believe the possibility that dogs were used in this context needs to be carefully considered. In my own frame of reference, well-trained dogs around livestock
are a definite asset, while poorly trained dogs are a liability. A dog or
dogs that cannot be controlled can disrupt an otherwise well-planned
hunting episode. However, a dog can be useful in recovering wounded
animals.

Gatecliff Shelter in central Nevada produced skeletal remains of several mountain sheep (Thomas and Mayer 1983). One concentration of
sheep bones from Horizon 2 at Gatecliff, radiocarbon-dated at about
1250 C.E., represents more than twenty animals. Seasonality data are
not as clear as might be hoped because of problems in interpreting rings
in teeth, though five teeth suggest that the animals were killed between
late fall and early spring. In addition, according to the investigators, the
location and topography suggest that the hunters could have used strategies of interception or ambush in the vicinity of the site. Being unfamiliar with the terrain, I am not sure that that I can clearly visualize the distinction between the two concepts.

David Thomas (1983a: 42–48) presents a well-informed discussion
of Nevada sheep in Monitor Valley. My experience with desert sheep has
been limited to short encounters, but as one might expect (and as
Thomas's accounts and those of several others who have hunted desert

sheep rams confirm), it appears to me that the general behavioral patterns of desert bighorn sheep and the Rocky Mountain sheep are quite similar. Like the latter, desert bighorns migrate seasonally from summer to winter ranges, and Thomas (1983a: 45–46) presents a case for a "migration-intercept strategy" of procurement as winter storms concentrate the animals and force them down traditional migration trails to winter quarters. Rocky Mountain sheep are most vulnerable when located on their winter range, and the same is probably true with desert bighorns.

Archaeological data are too limited to confirm aboriginal seasonal hunting strategies in the Gatecliff Shelter area. No mention is made of sheep traps such as those described in northwest Wyoming and adjacent areas, although I believe they would function as well with desert bighorns. However, such traps do require the presence both of building materials and of sufficient animals to justify the work needed to construct them, and those conditions may not have existed in Nevada. But when one is hunting sheep or any other animal, alternative procurement strategies are possible.

Perishable remains of past animal trapping complexes constitute a valuable source of information about prehistoric and historic life that is rapidly disappearing because of deterioration and forest fires. Mrs. John Murdock, a member of a pioneer family in the Dubois, Wyoming, area, remembers a sheep trap in the Whiskey Mountain area that burned during a forest fire in 1931. Before the fire, she helped salvage a large mountain sheep skull partially embedded in the fork of a pine tree; it is now part of a permanent exhibit in the Dubois museum. Many traps were lost in the enormous 1988 fires that burned off a large part of Yellowstone National Park and surrounding areas, and still others were lost in the lesser fires of 2000. In fact, only a last-minute desperate effort by firefighters managed to save the first one shown to me in the Sunlight Basin area by Roy Coleman. All are threatened further because their age has brought them to critical stages of wood rot. Conical and cribbed log structures made by the same cultural groups that made the traps are also threatened.

Lines of stones, fragments of pitch-impregnated logs, and isolated cairns at and above the timberline are most likely remnants of sheep-trapping complexes that appear much older than the ones with well-preserved wooden components. I am convinced that stone lines and horseshoe-shaped stone piles extending over a distance of 1.6 kilometers along a ridge at the timberline high above the North Fork of the Shoshone River in the Absaroka Mountains are the remnants of a complex designed to contain

FIGURE 40. Game blind just above the timberline in northwest Wyoming. (Photo by author.)

the animals within certain limits until the converging drive lines nearly come together at the edge of a steep talus slope. Located below are pieces of pitch-impregnated stumps and logs, suggesting the former presence of a holding structure or catch pen. The horseshoe-shaped rock cairns placed at several locations along the drive lines could be evidence of shamanistic activity or could be strategic spots for hunters to monitor the movement of animals and prevent their escape before they reached the end of the drive. This procurement complex is known as the Boulder Ridge sheep trap (see Frison 1991b: 248–49).

The Indian Point trap complex, above the timberline but in a different kind of terrain, lies high above the Wiggins Fork of the Wind River, and is another of the Dubois group (see map 7); it consists of lines of rocks and doughnut-shaped blinds (figure 40) shallowly excavated into the ground and built up around their edges with loose rocks and timber. As in the previous case, fragments of logs and stumps in a basin-shaped depression at the end of a natural gap through a narrow ridge suggest the remains of a catch pen; it appears to have burned at an unknown time in the past. Though today any source of timber is some distance away, at an earlier time there might have been timber available that was closer. A collapsed pile of stones mixed with pieces of timber on a promontory within view of the entire complex, but apparently not a part of it, has the earmarks of a shaman's structure. Unable to locate any time-specific artifacts, I have no reliable clue to its age; but the condition of wooden parts leads me to believe that it dates to within the past 500 years.

A PALEOINDIAN-AGE MOUNTAIN SHEEP NET

The North and the South Forks of the Shoshone River both originate in the Absaroka Mountains east of Yellowstone National Park; they join just before they enter the Big Horn Basin in northeastern Wyoming. Sheep Mountain, aptly named and with an altitude of 2,440 meters, is situated in the V formed by the confluence of the two rivers in what is presently part of a wintering area for mountain sheep. In the late 1970s, hikers found a bundle of cordage partially covered with and well-preserved by pack rat feces in a dry limestone alcove near the top of the mountain (figure 41). On close inspection, the mass proved to be a net with cordage made of juniper bark.

The net is too fragile to be more than partially unfolded, but it is thought to be between 50 and 65 meters long and between 1.5 and 2 meters wide. It was made of an estimated 2 kilometers of two-strand, S-twist cordage ranging in diameter from 1.0 to 5.2 millimeters. The mesh gage varies from about 0.71 centimeters to just over 3.1 centimeters. Well reinforced by the large-diameter cordage, it is entirely adequate for net trapping mountain sheep. It was folded along its length in a manner similar to that of a modern tennis net; top and bottom edges were doubled over to meet in the middle and then folded over once more. Finally, it was folded repeatedly, beginning at one end and ending at the other. Two wooden shafts were incorporated into the net, and charcoal from the burned end of one yielded a radiocarbon date of about 8800 B.P. (Frison, Andrews, et al. 1986), putting it in the Late Paleoindian period. Except for an occasional mule deer, no other large animals are known to frequent the immediate area (figure 42). I believe the net was stored there for seasonal use to trap sheep when they moved to their winter range. Judging from the observed behavior of mountain sheep when entangled in a net, two or three persons could hold such a net upright across a path while others drove animals into it. Once entangled, the animals could easily be (and most likely were) killed with clubs. The net could easily have been shifted to a different area, or other nets could have been stored at other locations.

A net procurement strategy could help explain the scarcity of stone projectile points of the same age in these areas. Mummy Cave—a deep, stratified cave site located along the North Fork of the Shoshone River

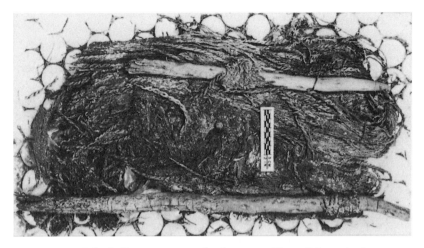

FIGURE 41. Paleoindian-age mountain sheep net. (From Frison, Andrews, et al. 1986: 355.)

a short distance from the east entrance to Yellowstone National Park, with a series of cultural levels dating from about 9,000 years to 400 years ago—produced one component about 8,000 years old that yielded three stone slabs placed on end along with a large ram skull, strongly suggesting ritual treatment. A fragment of a net similar to that from Sheep Mountain was recovered in a Late Archaic level in Mummy Cave, and sheep bones were recovered in all twenty-nine recorded cultural levels (Husted and Edgar 2002).

The mountain sheep trapping complexes in northwestern Wyoming described above probably were present in a much larger area; indeed, the appearance of the LaMarche game trap in southwestern Montana, at an elevation of more than 2,400 meters, is strikingly similar. The investigator, James Keyser (1974), believes it was designed for deer or mountain sheep, or possibly both. Given its configuration and location, I believe sheep to be the most likely candidates. Further west are two recorded game traps near the Idaho-Montana line at about 2,100 meters that, from their descriptions, are more difficult to interpret. The investigator, Bonnie Hogan (1974), suggests mountain sheep as the most likely candidates, with deer another possibility.

James Benedict (1996) performed an exhaustive study of high-altitude game drives in Rocky Mountain National Park in north-central Colorado. Most of the evidence is at or above the timberline and consists of cairns, circular blinds, lines, and walls, all of stone. Characteristic flaked stone

FIGURE 42. Location where the mountain sheep net (figure 41) was discovered. (From Frison, Andrews, et al. 1986: 354.)

artifacts suggest that these features were utilized for several thousand years, but it is difficult, if not impossible, to relate any particular set of stone features to specific hunting episodes, or to narrow the time frames of their employment as can be done with the better-preserved structures in northwestern Wyoming. Benedict believes that these stone structures may have been placed in strategic spots to take advantage of animal migrations of either deer or elk. Lacking an intimate knowledge of these structures, I have to defer to Benedict's interpretation of their past use.

Most investigators believe the late Pleistocene to early Holocene mountain sheep *(Ovis catclawensis)* to have been slightly larger than the present species *(Ovis canadensis)*. Sheep remains are rare in Paleoindian sites, although a ram skull but no other skeletal elements appeared in a Folsom level at the Hanson site in northern Wyoming (Frison and Bradley 1980). Unfortunately, it was too badly deteriorated to allow comparison with known *O. catclawensis* skulls. Skeletal remains have appeared in several Archaic-age sites but lack associated information that might reveal past hunting strategies. The preserved wooden sheep traps described above demonstrate a thorough knowledge of sheep behavior and provide us with our best data on which to base our interpretations of past communal procurement strategies, while the preserved net provides some insight into another method of communal hunting used during Late

Paleoindian times. Overall, the archaeological evidence in the Greater Yellowstone Ecosystem, although relatively limited, indicates continuity of mountain sheep procurement for more than 10,000 years.

Die-offs occur regularly and can rapidly impair the viability of herds. These result from numerous conditions, including parasites introduced by domestic sheep and pneumonia brought on by stress due to extreme winter weather conditions, malnutrition, and excess pressure by humans. In fact, pneumonia killed an estimated 400 of the Whiskey Mountain sheep herd at Dubois, Wyoming, during a single winter in the late 1990s. The herd, once considered to be one of the healthiest in the Rocky Mountain area, had not recovered sufficiently by 2002 to allow trapping for transplant purposes. Similar die-offs surely affected animals in earlier times, possibly accounting for some of our lack of past evidence.

Mountain sheep demonstrate behavioral patterns quite different from those of deer or pronghorn, and hunting strategies that are successful with sheep will not necessarily be effective for other animals. A constant threat to mountain sheep populations is their inability to successfully adapt to humans' presence and to build up immunity to parasites and diseases introduced by domestic sheep.

CHAPTER 7

Hunting Deer, Elk, and Other Creatures

MULE DEER

Rocky Mountain mule deer *(Odocoileus hemionus)* and white-tailed deer *(Odocoileus virginianus)* belong to a genus believed to have originated in the Old World and thought to have arrived in North America more than 2 million years ago, though the fossil record is far from clear. While the two species are distinct, their ranges overlap; and at the present time at least, interbreeding does sometimes occur and produces aberrant pelage coloring and antler forms. Mule deer are an adaptable species and their habitat range includes most of the western United States and Canada (see Wallmo 1981: 3).

> The multitudinous habitats of the mule and black-tailed deer subspecies are so diverse as to defy generalization. Rocky Mountain mule deer occur in the entire gamut of vegetation types, from tallgrass prairies on the eastern side of their range, westward across shortgrass plains and all shrublands, woodlands, and forest types of the Rocky Mountain region, to the desert scrub of the Great Basin. The southern end of their range is hot, arid desert land; the northern end is underlain by permafrost. (Wallmo 1981: 10)

Rare indeed is the present-day large animal hunter in the western United States who did not acquire a large share of his or her early field experience while pursuing deer. In my own time as a novice hunter, bringing home the first deer was a rite of passage of some importance. The word rapidly circulated through the community and the successful new hunter was accorded a noticeable increase in status, particularly if he or she could present evidence of a clean kill of a buck with a good set of antlers. In addition, of course, the success was a valuable addition to the food supply for the coming winter months.

In contrast to pronghorn, mule deer generally prefer areas of rough topography, steep slopes, and moderate vegetative cover. They are gregarious animals: groups of five to up to thirty or more are common, and the larger groups usually include a few young males displaying spikes and forked horns. A doe, her fawn, and occasionally a yearling are also a commonly observed small group. Mature males usually form separate groups of as few as two to as many as twenty for most of the year except for the rut, which begins in late fall. As antlers harden during the summer months, their loose velvet is removed by being rubbed on tree branches and brush, leaving loose strips of bark and exposed sap wood as unmistakable evidence of a male's presence. As the rutting season approaches, their necks swell noticeably and they acquire an objectionable odor that can carry over to and detract from the taste of their meat, especially toward the end of the rut. The experienced deer hunter always remembers to remove the scent pads located on the insides of the rear legs.

Mature males are more cautious than does and fawns; they usually allow the latter, along with the immature males, to cross open spaces while they keep to the rear and take advantage of brush, timber, and other obstacles to remain less visible. Inexperienced hunters seeking trophy males are often distracted by the movements of does and fawns and fail to notice larger males trying to remain out of sight. Does and fawns tend to move about and browse during the day, except in hot weather, when they move into shade. They seek sunny spots on clear winter days and thick brush or protected overhangs along cliffs during severe winter storms. They like to bed down in brush and timber patches and, in extreme cold, sometimes sleep in deep snow, which serves as protective insulation. On several occasions while riding for cattle in cold weather when there was deep snow cover, I have observed only the tops of their heads and large ears visible above the snow.

Snow depths of more than 40 centimeters can threaten the mule deer's survival. The grasses and other low ground-dwelling plants are covered, and deer do not paw through snow for forage. They then congregate, sometimes in large herds, and "yard up" in flat areas—usually along a stream or dry arroyo, or in another area where they can browse on brush and tall grasses. Once they have consumed whatever vegetation is exposed, fenced in by the deep snow and weak from lack of forage, starvation is inevitable unless a chinook wind or winter thaw exposes more vegetation. Under these harsh conditions they are vulnerable to predators; mountain lions in particular are highly efficient killers of mule deer, and coyotes are almost as serious a threat.

Mature males prefer to bed down early in the morning on high spots along ridgetops and steep slopes that provide good visibility for detecting predators both animal and human. Unless sick or weak from old age or long exposure to deep snow and shortage of food, mule deer seem better able to cope with livestock fences than pronghorn, though they too sometimes die while attempting to jump fences. Mule deer are not regarded as difficult animals to hunt. A fatal flaw in their behavioral pattern is that when disturbed, most tend to run a short distance and then stop to look back. But while does, fawns, and young males are easy prey, older males become surprisingly adept at avoiding human predators. During the day, they choose places where by sight and scent they can detect the approach of humans and other predators. In addition, their chosen location almost always has more than one route for rapid escape. Matching wits with old and wise bucks is a definite challenge. It is not uncommon to come across remains of deer that have died of old age or were killed by some animal predator, but not by a hunter's bullet.

Much of mule deer winter habitat is in areas within sight of mountain ranges and isolated uplifts that allow summer and winter migrations, sometimes over distances approaching 100 kilometers. Other mule deer prefer to remain in the rough country, commonly referred to as breaks, along streams. In fact, mule deer may be found unexpectedly almost anyplace with adequate browse and protective cover. Mule deer abandon much of their fear of humans in their pursuit of the first spears of green grass in spring. On more than one occasion after an especially bad winter, I have seen mule deer that have collapsed and died in desperate pursuit of that earliest grass. Severe spring weather, a not uncommon occurrence, can eliminate the older and weaker animals. However, the ones that do survive rapidly recover their strength; and as new grass becomes plentiful, they follow its growth to the higher elevations of their summer ranges, some of which can be at the timberline and above. The birthing period is in May and early June. Twin fawns are common, allowing fast recovery of deer populations following large die-offs. The animals lose their heavy dark winter coats in the spring and acquire light, dull red coats for the warm months. Summer feed is critical: during this period they must acquire the fat reserves needed to prepare them for the coming winter.

The first major storm in the fall, generally in late October or early November, usually triggers the beginning of the migration back to winter range. The fall migration is sometimes gradual; sometimes, when a severe storm of several days' duration brings deep snow and cold temperatures, the deer pour out of summer high country locations in large herds. Both spring and fall migration routes are well established, and most likely have been long used in the past. Present-day deer hunters take advantage of these predictable movements; prehistoric hunters were undoubtedly aware of them too and turned them to their advantage.

Mule deer provide a dependable source of food if herds are not overexploited. Mule deer are larger than pronghorn. A juvenile field-dressed male usually weighs about 50 kilograms, a weight that increases up to as much as 90 kilograms for a mature male in his prime; does weigh about 20 percent less. A field-dressed mature male mule deer in rough country and far from camp is too heavy a load for the average hunter to carry in. It is better to halve a large carcass and carry that rather than risk contamination of the meat while dragging an animal in dry or wet weather (though the problem can be avoided when there is good snow cover). According to Theodore White (1953a: 397) a properly butchered and cared-for mule deer carcass yields as much as 45 kilograms (100 lbs.) of meat. Provided the animal is healthy and in good bodily condition, the meat will have good flavor and high nutritional value.

Green pastures during spring, summer, and fall and haystacks in winter attract deer; if present in large enough numbers, they can consume significant amounts of forage intended for use by domestic animals, thereby creating friction between private landowners, hunters, and game management personnel. These troublesome deer are usually controlled through special hunting seasons and high fences to prevent access to stored hay.

MULE DEER IN ARCHAEOLOGICAL SITES

Mule deer remains occur frequently in plains and mountain archaeological sites (map 8), but usually only in small numbers; only in rare cases do they dominate the faunal assemblages or reveal evidence of communal procurement such as that documented for bison, pronghorn, and mountain sheep. Given the relative ease of mule deer procurement, the scarcity of their remains in these sites is puzzling and remains unex-

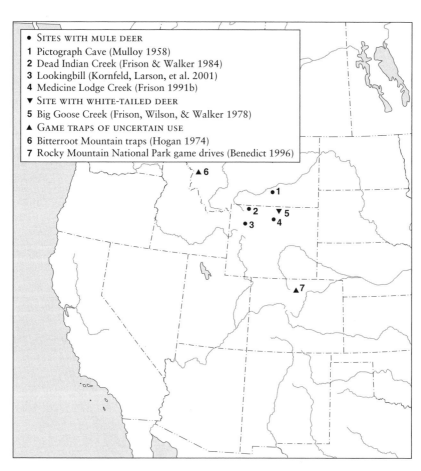

MAP 8. Selected archaeological sites with deer remains.

plained. Their presence in Paleoindian sites is minimal: for example, partial remains of two deer, including part of an antler, were recovered from the Horner site of Cody Complex age in northwestern Wyoming (Walker 1987: 336–38). Partial remains of one and possibly two mule deer were reported from the Jurgens site, also of Cody Complex age, in northern Colorado (Wheat 1979: 30–31). The Goshen-Folsom and Agate Basin levels at Locality I at the Hell Gap site each produced remains of a single mule deer; the provenience of other deer remains from Paleoindian levels is in doubt because some of the excavation data were lost (Rapson and Niven 2002).

Known foothill and mountain Paleoindian sites also demonstrate a scarcity of mule deer remains. Each of three Late Paleoindian levels dated between about 8,000 and 9,500 years ago in Mummy Cave, located in the Absaroka Mountains in northwestern Wyoming, produced partial skeletal remains of a single deer (Harris 2002). A cultural level dated at about 9,500 years ago at the Medicine Lodge Creek site (see map 8) in north-central Wyoming produced several broken and complete long bones and foot bones of a mule deer (Walker 1975: 88). The same level produced a stone-flaking hammer made from the proximal end of a large deer antler (Frison 1982e). Several cultural levels in Pictograph Cave (Mulloy 1958), a deep, stratified site near the Yellowstone River in southern Montana, produced mule deer remains (Olson 1958).

Altogether the plains and mountain archaeological records reveal little about prehistoric mule deer procurement strategies, but two sites are worth mentioning. The Dead Indian Creek site (see map 8)—located in what is known as the Sunlight Basin, a small basin completely enclosed by the Absaroka Mountains in northwest Wyoming—produced a large quantity of mule deer remains along with indications of some sort of associated ceremonialism (Frison and Walker 1984). Remains of at least fifty mule deer were recorded, and much of the site is still intact. Dental analysis demonstrates that the deer were killed and introduced into the site during the winter months (T. Simpson 1984). The site is of Middle Archaic (McKean) age—that is, it was used between 4,000 and 4,500 years ago—and the occupants were living in pit houses, one of which was over a meter in depth (see Hilman 1984: 41–45; Frison 1991b: 100). A depression adjacent to the pit house approximately 0.6 meters in depth contained several mule deer skull caps with attached antlers. One was placed in a depression in the center of a rock cairn and others were placed to the side (figure 43). One set of these antlers would be ranked high in the Boone and Crockett trophy class today. Smaller numbers of mountain sheep remains were recovered (Fisher 1984: 73–77), but not enough to determine the season of their death. Mountain slopes adjacent to the Dead Indian Creek site are covered with scattered to thick stands of timber and brush along with bunch grasses, a favorable setting for both mule deer and mountain sheep in winter. An interesting contrast to this location is provided by another Sunlight Basin winter occupation, known as the Bugas-Holding site, which has bison and mountain sheep remains but no evidence of deer; it is a more recent occupation, however, dated at about 400 years ago (Rapson 1990).

The Lookingbill site (see map 8), at an elevation of 2,621 meters and located in the southern Absaroka Mountains in northwestern Wyoming, contains several stratified Early Plains Archaic levels. An exposure of

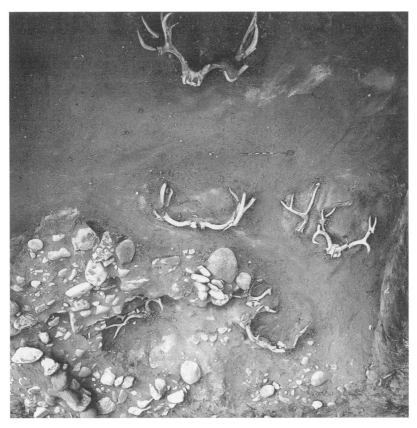

FIGURE 43. Arrangement of mule deer skulls at the Dead Indian Creek site. (From Frison 1978: 272.)

several square meters in one level revealed a concentration of skeletal re-mains representing several mule deer that, based on their tooth eruption and antler development, were killed during the summer (Kornfeld, Larson, et al. 2001). The level is undated, although a radiocarbon date on one of the Early Plains Archaic levels, 7140 B.P ± 160 years. (RL-554), is probably close to the time of the deer remains. Fortunately, more of the deer bone level is still intact for future excavation. At present, this is a favorite area for deer to spend the summer, with its wide expanses of open grasslands, tall sagebrush, scattered timber, and canyons with steep, rocky slopes. Deer need only move a short distance to find lower elevations along stream and arroyo bottoms with less harsh weather conditions and adequate feed with which to survive the winter. The site is in the immediate vicinity of several of the mountain sheep traps described in chapter 6, though few of their remains were recovered in any of the site components. But because mountain sheep tend to summer at slightly higher elevations, they may not have been available for hunting during the season when the site was occupied (see also Frison 1983).

WHITE-TAILED DEER

White-tailed deer are smaller in size and behave differently than mule deer. They are less migratory and seldom move far from a relatively small area throughout their lifetime. They prefer brushy areas, especially river bottoms, and rely more on hiding than mule deer do. According to Valerius Geist,

> The hiding strategy is far better developed in whitetails than in mule deer. Its success depends on the deer's remaining rigid and silent. It also requires a quick getaway when the deer is surprised, hence the explosive eruption from hiding, the very fast gallop low over the ground, and the avoidance of obstructions. To maximize time for decision-making while escaping and thus maximize distance between itself and the predator (as well as diffusing its scent), the white-tailed deer requires a trail system. In tall grass and thickets, as well as in snow, a trail system is a necessity. Unlike the mule deer, the whitetail minimizes contact with obstructions to running. Once some distance is gained, the whitetail stops, normally in cover. (Geist 1981: 172)

Whitetail hunting strategies can change rapidly with changes in vegetative cover. When the leaves fall and the ground vegetation dies in the fall, they become much more visible. Better yet, postpone the hunting until the first good snowfall, which serves a double purpose: it increases animal visibility still more and decreases the noise of the hunter's moving through dried leaves, grass, and underbrush. If they believe they are undetected, hidden whitetails will often allow a person to pass by within a few meters. I remember once standing atop a pile of dead tree limbs stacked for burning when I suddenly realized I was looking at the tops of a large set of antlers and not tree limbs, well into the pile and less than three meters away. I moved to within a meter of the antlers before a male whitetail, with the largest set of antlers I've ever seen, exploded from the edge of the brush pile and was soon out of sight in thick timber. Whitetails also lack the bounding or stotting gait found in mule deer; and when flushed out of hiding, they demonstrate remarkable speed and an ability to keep trees, brush, and other obstacles between them and the hunter. When those who hunt both species gather, it is easy to start an argument over which is the more difficult to hunt and which is the best eating. Having had long experience hunting both species, I call it a draw on both counts.

The plains and Rocky Mountain archaeological records reveal even less on prehistoric white-tailed deer procurement than on mule deer. A single white-tailed deer was recorded at the Lindenmeier Folsom site in northern Colorado (Wilmsen and Roberts 1978: 46) and possibly another one at the Cody Complex Jurgens site, also in northern Colorado (Wheat 1979: 31). The 450-year-old campsite adjacent to the Big Goose Creek Bison Jump (Frison, Wilson, and Walker 1978; see map 8) yielded a large number of nearly complete white-tailed deer skulls, antlers, antler tool handles, and antler parts discarded in manufacture. Since most white-tailed and mule deer antlers are easily identified, this find leaves no doubt as to the presence of white-tailed deer in northern Wyoming in Late Prehistoric times, though they were not reported there historically until early in the twentieth century.

Bengt Anell (1969: 24–32) argues for a much greater emphasis on deer along with a wide assortment of deer procurement strategies used by Native North Americans in the California, Northern Plateau, and Northwest Coast areas. These strategies include using nets, snares, pitfalls, enclosures, water drives, fire drives, surrounds, fences with snared openings, and ambushes in natural features; building temporary enclosures on thick ice; and driving animals onto thin ice and over cliffs. He also claims that in certain instances hunters used trained dogs, made noise with artificial rattles constructed of deer mandibles hung from tree branches, imitated wolves' howling, and came together in large numbers to beat the bushes in order to frighten the animals into different kinds of containment features. There is a significant amount of overlap between the occupation areas of white-tailed and mule (or black-tailed) deer in western North America (see Hall 1981:1090, 1094), but Anell makes no distinction between the two, though their behavioral differences would have required somewhat different procurement strategies. Anell relies on accounts by reliable ethnographic observers; however, the presently known archaeological and ethnographic records fail to produce satisfactory evidence in the plains–Rocky Mountain areas of the kinds of prehistoric deer-hunting strategies he names.

Because the three areas on which Anell focuses are mostly outside of good bison habitat, one would expect deer, if present in sufficient numbers, to have taken on more importance as a food resource and consequently to have been the targets of serious procurement efforts. In the

Northwest Territories, where deer and caribou habitat overlap, Anell claims that the same procurement features, particularly snare corrals, are used for both (see Anell 1969: 15–21). I believe there is some confusion in Anell's account, particularly in reference to the Chippewa. However, caribou *(Rangifer tarandus)* and mule deer habitats do overlap in Chippewa territory (see Hall 1981: 1090, 1105; Murdock 1960: 35); it is possible that some caribou procurement strategies and features might have also sufficed for deer.

I am unable to claim other than a nodding acquaintance with caribou, although on one occasion I had the opportunity, just north of the Brooks Range in Alaska, to view from a helicopter a caribou procurement complex from all angles. In this complex, a line of stone cairns set nearly 10 meters apart and estimated to be a kilometer in length led to a lake where, as was graphically recounted to me, the swimming animals were killed by hunters in canoes. One arm of the lake was situated so that the hunters in canoes were out of sight of the animals until they were well into the water. The site was apparently prehistoric; no historic or perishable items were present.

It has been suggested that two game traps reported in the Bitterroot and Castle mountains in western Montana (Hogan 1974; see map 8) could have been used for deer or sheep. However, as I noted in chapter 6, their location and configuration strongly suggest to me that they were most likely meant for sheep.

ROCKY MOUNTAIN ELK

At present, the Rocky Mountain elk *(Cervus elaphus nelsoni)* is well established throughout the Rocky Mountain areas in the United States and Canada. Other subspecies—the Manitoban, Tule, and Roosevelt elk—occupy smaller areas (Bryant and Maser 1982: 25) and need not be considered here. Early western travelers, especially Lewis and Clark, gave the impression that there were large numbers of elk, more at home on the plains than in the mountains (see M. Lewis [1904–05] 1969: 4.40–41). However, elk suffered much the same harm as bison from market and trophy hunting in historic times; though eliminated from most of their original habitat, they were fortunate to have adapted well to a variety of other environmental settings. Thus many were able to find refuge in the inaccessible areas of Yellowstone National Park and Jackson Hole in northwestern Wyoming.

During the late 1800s and early 1900s, an estimated 80 percent of the elk in the United States were in the Yellowstone National Park and Jackson Hole areas. In 1913, the National Elk Refuge was established in Jackson Hole; of the estimated 70,000 elk remaining in the United States in 1919, about 65,000 were in the Yellowstone–Jackson Hole areas. Subsequently, these two sources provided almost all of the animals for transplants needed to reestablish herds in locations where they had been eliminated earlier (see Robbins, Redfearn, and Stone 1982).

Elk canine teeth, also known as buglers, ivory teeth, whistlers, or tusks, were prized as decorative items by Native Americans. When elk populations declined and canine teeth became scarce, imitations were made from bone (see McCabe 1982:105–10). Unfortunately for the elk, white men also cultivated a desire for elk canine teeth, and tusk hunters contributed to the decline of elk populations during the late 1800s and early 1900s. Pairs of elk canines properly mounted became an unofficial membership badge for members of the Benevolent and Protective Order of Elk (B.P.O.E); depending on their size and quality, a pair would bring from $10 to $20. The result was the killing of large numbers of elk for their canines only, with the remainder of the carcasses abandoned. Thankfully, the practice became less common after the B.P.O.E. condemned it (see Potter 1982: 542–44).

My own earliest experience with elk was with herds that frequented the higher elevations of the Big Horn Mountains in northern Wyoming, usually on both sides of the timberline in summer and in the foothills during the winter. Ridgetops with grass exposed by the wind and steep slopes with browse plants such as curlleaf mountain mahogany (Cercocarpus ledifolius) are favored wintering habitats. Within the past three decades, herds have been successfully reestablished in much of the Wyoming Basin, where they remain throughout the year. Limited hunting is allowed, and the elk have become particularly adept at hiding from hunters in areas of moderate topographic relief and in the vast expanses of stabilized and active sand dunes, scattered junipers, and tall sagebrush that characterize the Wyoming Basin. This ability correlates well with earlier historic accounts of elk being a plains animal.

Rocky Mountain elk are gregarious. If undisturbed, herds of a hundred or more are common; but as hunting pressures increase, they tend to fragment into smaller groups and seek refuge in rough and timbered country. They have good eyesight and can readily detect human presence by smell. Elk also emit a distinctive scent: small groups tend to spend the day in thick patches of brush and timber, but even a slight air movement can alert a hunter to their presence. The best strategy I was able to devise for these situations was to scout out the leeward side of timber patches, locate the animals by smell, and then run at top speed into the herd. This nearly always confused the animals, most of whom were bedded down, causing them to charge blindly in all directions. Those unable to locate the rest of the herd would run first in one direction and then in another, paying little attention to my presence and usually making it possible for me to kill one or more of them. On the other hand, a hunter attempting to sneak up on a bedded-down herd is usually detected by the animals, who immediately charge off as a group, allowing only an occasional glimpse of their rear ends and no chance at a lethal shot.

Elk are good travelers, and under pressure will move with a running gait that rapidly covers long distances; they stop only occasionally to look back and check for pursuers. It requires a better-than-average saddle horse to keep pace with them when they run, and their long legs enable them to travel well in deep snow and thick sagebrush. A herd is always led by a mature female; if she is killed, the remainder of the herd usually becomes confused, often running short distances and reversing directions until another leader takes charge. I have more than once observed a running herd of elk, upon losing its leader, stop and mill around indecisively for a short while before moving on, all the while seemingly oblivious to the nearby hunters field dressing dead animals.

Ordinary livestock fences do not present serious obstacles to the movements of elk, although an occasional dead animal is found entangled in wire. I recall one instance when one end of a length of barbed wire about 300 meters long became entangled in the antlers of a large male in late November. He dragged the loose end throughout the winter and managed to survive until he shed the antlers in March of the next year. A closely packed herd of elk stampeded in winter by thoughtless people on snowmobiles can move an entire section of fence several hundred meters and even more. On the ranch, one of my less exciting duties every spring was to repair fences torn up by elk in winter.

Like mule deer, male elk congregate in groups of two to a dozen or more for most of the year. However, before and during the rut in late summer and early fall, dominant males exert their authority and any earlier modicum of camaraderie breaks down. Antler clashing and bugling now dominate their daily activities. I can think of nothing that arouses a hunter's desire to begin the hunt more keenly than the distant bugle of a male elk on a clear, cold morning in early fall. When a practiced hunter answers the call of a male elk, who then believes he is being challenged by another male, the animal will often abandon all caution and charge at full speed in the direction of the sound to drive off what he believes is another male threatening to take over his harem. Observing battles between mature male elk from a good vantage point is an experience not soon forgotten (see also Geist 1982). It is not always necessary to produce even a close imitation of an elk bugle; often even a shrill whistle will be answered. Under certain conditions, females emit a squeal that can be heard for short distances and often reveals the location of a herd hidden by brush and timber.

Elk are roughly twice the size of mule deer; a healthy, field-dressed mature female in good condition in late summer or early fall will tip the scales at about 160 to 180 kilograms, and males are at least 15 to 20 percent heavier. Their larger size affords them more protection than mule deer from predators such as mountain lions and coyotes. However, mountain lions are known to take down calves up to a year old. Grizzly bears are also known to kill elk calves. Management of elk herds in Yellowstone National Park is a perennial problem, and the recent introduction of wolves there is expected to aid in eliminating the excess numbers of elk. This is an experiment whose results all who are concerned with wildlife management are awaiting with great interest.

ELK REMAINS IN ARCHAEOLOGICAL SITES

The archaeological record for the plains and Rocky Mountains doesn't contain many elk remains except for antlers. The Folsom level at the Agate Basin site produced a tool made by deeply grooving into the mainshaft of a large elk antler and removing a short section of the mainshaft and the brow tine (figure 44). It is believed to have been used as part of a device to remove flutes from Folsom points (Frison and Bradley 1981). A broken and refitted elk antler projectile point and a proximal end of a similar one recovered among the skeletal remains of several bison and pronghorn in the same site component are described in chapter 8, which discusses weapons and tools. Two cut-off tips of elk antler tines, one believed to be the spur for an atlatl, were recovered from the Hell Gap cultural level at the Agate Basin site. However, no elk bones were found in any of the Agate Basin site components. The one elk antler tine believed to be an atlatl spur is described in the next chapter.

Excavations in the late Paleoindian Frederick cultural level at the Hell Gap site in the 1960s produced a stone-flaking hammer made from the proximal end of a large elk antler. The estimated date of the Frederick Complex is between 8,000 and 8,400 years ago (Irwin-Williams et al.

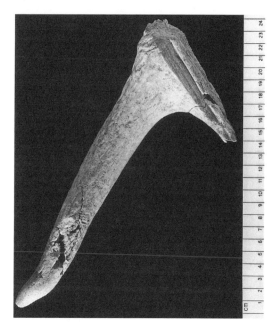

FIGURE 44. Elk antler tool from a Folsom
component at the Agate Basin site. (From Frison
& Bradley 1981: 13.)

1973) and again, no skeletal remains of elk were recovered. Two broken
elk metatarsals from the Cody Complex–age Jurgens site mentioned earlier demonstrate cut marks (Wheat 1979: 31).

Large numbers of elk antler digging tools (figure 45) were recovered
in an Early Plains Archaic–age tool stone quarry in the northern Big Horn
Mountains that was radiocarbon-dated at 6,200 B.P. ± 170 years (RL-
677; Frison 1991b: 262, 291, 292). Such tools were also found in the
large Late Archaic–, Pelican Lake–age Schmitt chert mine near Three
Forks, Montana, dated at about 2,200 years ago (Davis 1982). Skeletal
parts of elk would not be expected at the latter two sites, which I mention only to establish the presence of elk antlers used as tools. All of the
elk antler tools with which I am familiar through Archaic times appear
to be from shed antlers.

Elk remains appear with greater frequency in the Late Prehistoric
period, although antlers are still the parts most often represented. For example, the Big Goose Creek site (Frison, Wilson, and Walker 1978), the
Piney Creek site (Frison 1967b), the Bugas-Holding site (Rapson 1990),
and a Late Prehistoric–age locality at the Dead Indian Creek site (Jame-

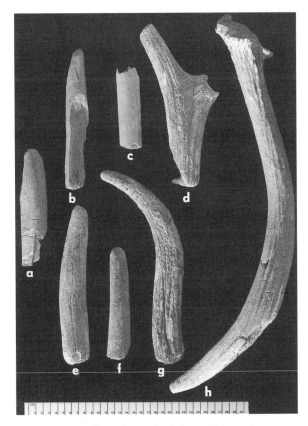

FIGURE 45. Elk antler *(a, b, d–h)* and bison rib
(c) digging tools from a stone quarry in northern
Wyoming. (From Frison 1978: 331.)

son 1984: 45–48) all produced tools and debris from the manufacture of
elk antler tools. The common L-shaped hide flesher made from part of
the main beam and the brow tine of an elk antler appears widespread in
Late Prehistoric or Early Historic times. After the appearance of the horse
on the plains, structural parts of frame saddles such as the "prairie chicken
snare saddle" (see Ewers 1955: 91–93) were made from both elk and deer
antlers, and a common style of quirt utilized part of an elk antler tine.

I mentioned earlier two large piles of elk antlers in Montana and another in northern Wyoming that very likely had some sort of ritual significance. George Grinnell refers to a location, somewhere to the west of the
Black Hills and probably in northeastern Wyoming, "where in early times
the Arapahoes drove elk over the cliff, and after they had finished killing

and caring for the meat, it was their practice to carry the horns [antlers] of the animals to a great pile of elk-horns already heaped up and add them to it, so that at length there was an immense pile of these antlers" (Grinnell 1923: 1.276). On the same page Grinnell states, "It is altogether probable that all the Plains Indians—certainly Arapahoes, Blackfeet, and Cheyennes—captured elk by this method." The events he recorded occurred after the Plains tribes had acquired horses, and they need to be confirmed by the archaeological record. Unfortunately, despite the relative recentness of these events, the evidence, in the form of the antler piles, has had sufficient time to rot away.

Stuart Conner of Billings, Montana, an expert on Montana history and archaeology, has collected accounts from Native Americans and from the journals of early western travelers; from them he has concluded that the Montana elk antler piles were most likely of Crow Indian origin and probably had some religious significance (Conner 1970). The elk antler pile in the Big Horn Mountains of northern Wyoming, mentioned above, was located in Crow territory and very likely also is attributable to them. Still found occasionally are large male elk heads and individual shed antlers in Prehistoric or Early Historic lodges (figure 46), in trees, and in protected locations. These too, though the conjecture certainly remains unproven, may have had ritual significance similar to that of the sheep heads in trees near mountain sheep traps, as described in chapter 6.

A Late Prehistoric cultural level, almost certainly Shoshonean, at the Dead Indian Creek site in the Sunlight Basin in northwest Wyoming yielded several elk bones along with a proximal end of a large elk antler that displays deep grooving, presumably made as antler strips were being removed in the process of manufacturing tools (figure 47). In this case the antler was attached to the skull (Jameson 1984: 47), indicating that it had not been shed.

Elk antlers now represent a cash crop. The Boy Scouts in Jackson, Wyoming, have a special use permit to gather the shed elk antlers annually from the National Elk Refuge. This is a one-day affair and, on one recent occasion, about 1.8 metric tons of antlers were gathered. Most of these are auctioned off to buyers from Korea and Japan, where they are shipped and then cured and ground into powder, which is sold as an aphrodisiac. Some of the proceeds of the auction sales are used for Boy Scout projects; the profits also aid the elk-feeding program on the refuge during the winter (see Potter 1982). With elk antlers bringing $3.00 and more per kilogram, and with the antlers from a healthy mature male usually weighing in the neighborhood of 9 kilograms, areas where large numbers

FIGURE 46. Elk antlers from a prehistoric lodge in northwest Wyoming.
(Photo by author.)

of elk shed their antlers in the spring have become the scenes of stiff competition among antler gatherers. I know of one case in which eighteen mature male elk spent the winter in a secluded location. They were monitored from a distance through a spotting scope, and their antlers were
collected immediately after being shed. The sale of the antlers brought more
than $500.00 in 1968, a significant amount of added income at that time.

It is illegal to collect antlers from Yellowstone National Park, but the
prospect of easy money often proves irresistible. One strategy is to collect a large quantity of antlers, load them on a raft, and attempt to float
them down the Yellowstone River and so out of the park. The perpetra-
tors are usually apprehended by park rangers; some slip by but fail to
take into account the rough waters in the Yellowstone River that then
capsize their rafts and dump the antlers into the river. On at least one
occasion, the rafters drowned.

Deer and elk have not dominated the plains-mountains archaeological record as have the mammoths, bison, pronghorn, and mountain sheep.
We do not know whether that record truly reflects the economic value
that prehistoric inhabitants placed on these animals. They were present
throughout the prehistoric periods, although in what numbers is not clear.
Procurement strategies for each are easy to develop. Their absence in the
archaeological and anthropological record as we understand it may mark
their actual absence from humans' daily life, or it may signify that our
understanding is inadequate.

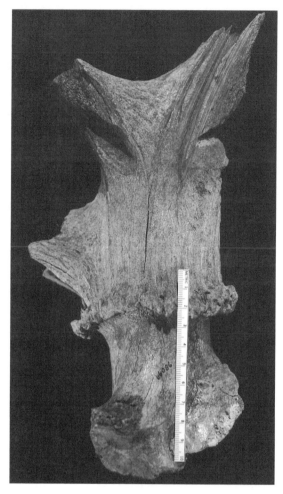

FIGURE 47. Base of a large elk antler with deeply
incised grooves. (From Jameson 1984: 47. Photo
by D. N. Walker.)

BLACK AND GRIZZLY BEARS

Black bears are found throughout the foothills, mountains, and other secluded areas that provide adequate protection. They present relatively
little danger to humans, although often, especially when their food sources
are low, they wander into populated areas where human emotions pre-
vail and law enforcement and wildlife officials are called to dispatch or
to trap and relocate them. As mentioned in chapter 1, I have had little

desire or reason to kill a bear, though I have been in many situations when it would have been easy to do so. I do recall one tense moment when, not paying close attention, I realized I was between a female and her cubs. She immediately charged toward me—but as the cubs took off in another direction, she forgot about me and started running after them.

Bears are especially attracted to chokecherries *(Prunus virginiana)* and serviceberries *(Amelanchier* sp.), which ripen first at the lower elevations, then higher as summer progresses, making them available for several weeks. Nick Wilson (E. Wilson and Driggs 1919: 62–65) recalled an occasion when a black bear mauled a young Shoshone girl among a group picking berries. The remainder of the berry-picking group immediately abandoned the victim. When he asked why they retreated rather than try to help the victim, the father of the mauled girl was said to have replied that in the past "the bears killed many people because they [the people] tried to help the one that was first caught" (E. Wilson and Driggs 1919: 64). Although he made no distinction between black and grizzly, Wilson further claimed that the Indians adopted a hands-off approach to bears unless they were encountered in open country. On another occasion, when two black bears were spotted in the open, Wilson joined several other mounted hunters who pursued the bears and killed them.

Grizzly bears, once widespread in western North America (see Storer and Tevis 1978: 3), now have dwindled to a mere fraction of their earlier numbers and territory. In the lower forty-eight states, they are restricted to parts of Wyoming, Montana, Idaho, and Washington; without the protection provided by national parks and federal and state agencies, they would very likely have been eradicated. Grizzlies are about twice the size of black bears and have a distinctly different temperament. At present, they pose the greatest danger in sudden and unexpected close encounters with humans. Backcountry visitors are well advised to make enough noise to alert the animals to their presence. Grizzlies' sense of hearing is well developed, and many backpackers attach a small bell to their gear. When threatened by a grizzly, people should avoid sudden moves, stand their ground, and above all never attempt to run away. If attack appears imminent or actually occurs, the best strategy is to attempt to kill the bear or else—a choice that requires considerable courage—play dead and hope the animal will then lose interest and move on. In wilderness and national forest areas, where regulated hunting of other large animals is allowed, the intended victim may be armed; but in national parks, where carrying firearms and hunting are illegal, killing a threatening grizzly is not an option. Pepper spray is often carried by backpackers as a final defense. My contact with grizzlies has been limited mostly to sightings from vehicles and helicopters. Once, while hiking in Alaska, I spotted a grizzly with a cub digging for food. I gave her a wide berth, and she gave no indication she was aware of my presence.

Upon emerging from dens in spring, grizzlies seek out winter-killed carcasses—mostly elk, but also deer and bison. Lacking these they can pursue and kill animals weakened by winter conditions. As spring progresses, plant foods appear; and rodents, along with the newborns and very young (especially elk), become easy prey. Spawning cutthroat trout *(Oncorhynchus clarki)* in small streams draining into Yellowstone Lake provide some sustenance. Whitebark pine *(Pinus albicaulis)* nuts and army cutworm moths *(Euxoa auxiliaris)*, which are attracted to high-altitude flowering plants, are important late summer foods as grizzlies prepare for winter hibernation. Domestic animals are vulnerable, especially during the years when, for various reasons, normal food sources fail to materialize. Bear images appear as likenesses of their bodies painted, pecked, and incised in rock art (see figure 2; see also Mulloy 1958: 133; Hendry 1983: 181–82); they also appear as rectangular and oval forms, with lines representing claws on one end that are commonly identified as bear tracks (Francis and Loendorf 2002: 157). The latter may appear singly or as several in a series. These representations suggest that bears held some importance to Native American spiritual beliefs and ceremonies, such as the Crow Bear Song Dance (Lowie 1924: 356–60).

BEAR HUNTING PAST AND PRESENT

In Wyoming, there is a bear hunting season in the spring as well as in the fall. Most bear hunters I know were seeking trophies rather than food, even though, according to Theodore White (1953a), a black bear can yield as much as 200 pounds of usable meat. I mentioned in chapter 2 my own aversion to eating bear meat because of bears' fondness for consuming the flesh of animals in advanced stages of decomposition.

The archaeological record reveals very little about prehistoric bear hunting. Their remains, though usually present in archaeological sites, constitute only minor percentages of known faunal assemblages. Past relationships between Native Americans and the grizzly are perhaps best known from historical and ethnological accounts in California. Until the Europeans arrived in California, "the big bears rather than man dominated the scene, because the natives lacked adequate weapons and were afraid of grizzlies, whereas the bears had little to fear from any living creature" (Storer and Tevis 1978: 77). This unequal power relation probably held over the entire area of prehistoric bear occupation.

Firearms aided in making the dynamic of human and grizzly encounters more equal, although hunting them was still dangerous. In California, different nineteenth-century grizzly-human contests were devised by small groups of Spanish vaqueros who were expert in the use of the rawhide lariat, or reata, while mounted on well-trained horses. After the bear was immobilized by loops thrown around its legs, it was then strangled with one or more loops tightened around its neck. Such competitions, dangerous for both horses and men (Storer and Tevis 1978: 131–39), were definitely a sporting activity, though probably an outgrowth of attempts to control grizzly populations as the bears killed domestic animals at an alarming rate.

Yellowstone National Park and its surrounding areas (the Greater Yellowstone Ecosystem) have maintained a grizzly bear population because of hunting restrictions and a remoteness that limits human access. Grizzly hunting was allowed outside of Yellowstone National Park until 1975, when the rapidly disappearing animals were finally accorded endangered species status. Since that time, management of the bears in Yellowstone has reversed the trend toward eradication but has also generated far more than its share of controversy (see McNamee 1984: 89–111).

In one respect, these modern bears might throw some light on prehistoric bear hunting. John Craighead, Jay Summer, and John Mitchell (1995: 233–78) liken the aggregations of grizzlies at Yellowstone's garbage dumps (a practice now strongly discouraged) to descriptions from historical accounts of bears gathering at beached whale carcasses and concentrations of dead land animals in California; they postulate that large prehistoric bison procurement sites might have elicited similar behaviors. Given their acute sense of smell and their reliance on winter-killed animals such as bison and elk when they emerge from their winter dens, grizzlies and also black bears would almost certainly have been quickly attracted to the stench from a bison kill. However, if the fear and avoidance of grizzlies shown by Native Americans also prevailed in the more distant past, I find it implausible that these kill sites were prehistoric hunting grounds for grizzlies. Instead, I would have to argue that a wise prehistoric hunter would have avoided irritating a grizzly by contesting its claim. Partial remains of a grizzly were recovered at the Vore Buffalo Jump in northeast Wyoming (Walker 1980: 156) but there is no evidence to indicate whether any agency, human or other, was involved in its death. Until better evidence materializes, any discussion of the strategies of prehistoric humans in hunting grizzly bears must remain largely speculative.

SMALL MAMMALS

The large grazers and browsers were not the only ones taken by prehistoric hunters. In addition, they pursued small vegetarians including jackrabbits, cottontails, marmots, prairie dogs, pika, muskrats, porcupines, and beaver. Carnivore and scavenger remains constitute relatively small percentages of faunal remains in archaeological sites, though it is difficult to assess how important they were as food. The smell emanating from dead animals always attracts other animals such as bears, wolves, coyotes, badgers, bobcats, and skunks. Mountain lions have little interest in carcasses, preferring to kill on their own. As mentioned in chapter 2, the best method to attract a bear is to kill an animal such as an old horse, open it up to release the maximum amount of odor, and simply wait. Carnivores, scavengers, and some rodents that like to chew on bones were undoubtedly attracted to kill sites of larger animals because of their odor, which could be detected over long distances. Their presence among the faunal remains suggests some may have been killed by human hunters; others may have died as they fought among themselves over carcass remains.

Ethnographic and travelers' accounts confirm that dogs were of importance to the Plains Indians (see, e.g., G. Wilson 1924: 196–230; Wied-Neuwied 1906) and were still used as work animals even after horses were introduced. These dogs were apparently partially domesticated wolves, and thus to differentiate between the "dogs" and wolves from skeletal materials recovered in archaeological sites can be difficult. However, analyses of skulls reveal one modification strongly indicative of human association: the canine teeth of some specimens have been reduced to the occlusal surface of the incisors, a condition not observed in known wolf specimens. This operation was probably performed to protect both humans and other animals. Moreover, Gilbert Wilson (1924: 201) claims that male dogs were castrated to make them more tractable and less inclined to run away.

In 1980, six canid skulls were found nested together and buried in a pit feature in northern Wyoming; first and second bison thoracics were crossed above the tops of the skulls. Four of the skulls demonstrate depressed fractures, suggesting that the animals were deliberately killed. In addition, the canines of all six were reduced in the same manner as found in one skull among five from the Vore Buffalo Jump (Walker and Frison 1982: 129–32). These skulls are all within the size range of wolves; I mention them here to emphasize that there may be an explanation for the presence of wolf-size canid remains in archaeological sites other than the actual hunting of wolves.

Cottontails (*Sylvilagus* sp.) and jackrabbits (*Lepus* sp.) can provide significant amounts of food, and each species requires a different procurement strategy. There is no unequivocal archaeological evidence that the Shoshonean strategy of net trapping jackrabbits (Steward 1938: 38–39) extended north and east into the Wyoming Basin, although jackrabbits are present there today. They like to hide under a bush or in a badger hole during the day and emerge at dusk, sometimes roaming in groups of fifty or more. Individual animals disturbed during the day usually make several long jumps and then pause momentarily to look back, thereby presenting a good target.

Cottontails hide in brush or remain visible at or close to holes into which they can dive at any indication of danger. The bow and arrow is the ideal weapon for cottontails because the arrow shaft makes it difficult for the wounded animal to disappear into a hole. However, at a very early age I was shown a different method: a green branch with a forked end could be inserted into the hole until it touched the rabbit. A few twists anchored it firmly enough in the rabbit's fur so that the animal could be pulled to the surface. The same strategy does not apply to the prairie dog, since its body is covered with short hair.

Wood rats, or pack rats *(Neotoma cineria),* that maligned small rodent that irritates people by chewing leather and carrying small items to its nest, are found throughout the western United States and were trapped for food prehistorically. Apparently an important food source for the Gosiute in the Great Basin, they were taken by the use of small deadfalls more commonly known as figure-4 traps (J. Simpson 1876: 53). Wood rats frequent the cracks and other openings in canyon walls and are present in considerable numbers in many areas. According to Howard Egan (1917: 237), a common strategy was to set traps one day where signs of the animals appeared and collect the dead ones the following day. According to Theodore White (1953a: 398), a wood rat yields 0.31 kilograms (0.7 lbs.) of meat—a calculation that, from my experience with wood rats, seems too high.

The supply of wood rats in an area would diminish rapidly under such trapping conditions, requiring hunters to move to a new area within a few days to repeat the process. Very few persons have seen a wood rat

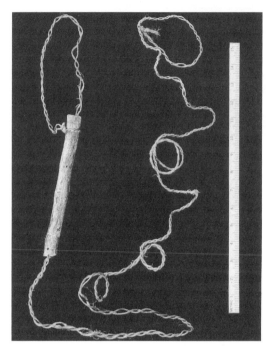

FIGURE 48. Small animal snare from a dry cave
in southwest Wyoming. (From Frison 1991b:
264. Photo by D. N. Walker.)

because they are mostly nocturnal and very skilled at staying out of sight.
Anyone unfamiliar with their habits would go hungry long before he or
she could devise a procurement strategy. Several small animal snares
(figure 48) have been recovered in dry caves in southwestern Wyoming.
I believe these would have been as reliable as figure-4 traps for obtaining wood rats.

Gilbert Wilson (1924: 165–67) described Hidatsa trapping gophers
(species not given) with a simple noose made from hair taken from a
horse's mane. The noose or snare was anchored to a stick and placed in
the hole close to the surface. When the gopher emerged far enough from
the hole, the noose was drawn tight around its neck. The hunters rapidly caught eight gophers, which were skewered and eaten. White (1953a:
398) claimed the same amount of meat is on a pocket gopher (0.31 kg)
as on a wood rat, although again, having personally trapped large numbers of gophers from hay fields, I believe that this figure is too high. Wilson also mentions drowning gophers out of their underground tunnels.

Beaver *(Castor canadensis)* and muskrat *(Ondatra zibethicus)* share an almost continentwide distribution north of Mexico and south of the Arctic. Both are found regularly in archaeological sites and are believed to have been used as food as well as for hides. The North American fur trade centered around the beaver. Beaver are regarded by many as pests because of their ability to eliminate trees, but they are also admired by many for their work ethic. Metal traps efficiently and dramatically reduced beaver populations in the late eighteenth and nineteenth centuries, during the fur trade years. The bow and arrow is an effective weapon for both beaver and muskrat.

Two closely superimposed cultural levels at the Medicine Lodge Creek site in the Bighorn Basin in northern Wyoming at the ecotone between the interior basin and the foothills produced a bone midden radiocarbon-dated at 9590 B.P. ± 180 years (RL-393). First believed to be an owl roost, tools and flakes but no distinguishing projectile points around the perimeter of an associated fire hearth finally led the investigator to conclude that it was a cultural component. The best-represented species present included at least 135 wood rats, 180 prairie voles *(Microtus ochrogaster),* 134 montane voles *(Microtus montanus),* 101 northern pocket gophers *(Thomomys talpoides),* 33 deer mice *(Peromyscus maniculatus),* 11 cottontail rabbits, and 5 jackrabbits. The least-represented species included at least one bison (a fetal bone), one mule deer, and two mountain sheep (Walker 1975). These numbers indicate a strong reliance on small mammals at a time when bison were the main source of food on the open plains. Whether small mammal procurement should be considered in the category of "hunting" is open to question. Yet it is undeniable that the strategies used here, like those used in large mammal hunting, require skills that come only with experience.

BIRDS

Remains of sage grouse *(Centrocerus urophasianus)* and to lesser extent blue grouse *(Dendragapus obscurus)* occur in small numbers in archaeological sites. The former thrive best in treeless, sagebrush- and grass-covered areas. They congregate in large bunches at certain times of the year, when they offer little challenge to hunters. In late summer they are attracted to water sources, where they are easily spotted and will take flight only if closely pursued. Blue grouse are found at higher elevations and are attracted to areas of scattered timber, especially near water sources. David Thomas (1983a: 57; 1988: 324–25) presents an interesting account of sage grouse procurement among Nevada Shoshoni that occurred in the spring at the same time as the strutting season. Nets were used to trap female birds, taking advantage of their tendency to move along the ground rather than take to the air. Barriers built of sagebrush were also used to herd sage grouse into brush enclosures (see also Steward 1941: 222).

Blue grouse are easier to obtain than sage grouse: about all that is needed in the way of a weapon is a heavy stick to throw at birds on the ground. If pursued closely enough, they usually fly to the lower branches of a nearby tree where they can be dislodged with rocks or sticks. The bow and arrow or atlatl and dart are the ideal weapons to kill birds that follow such patterns of behavior. As is the case with most small mammals and birds, the strategies needed to procure them are simple, and their pursuers thus can easily acquire the makings of a tasty meal.

In summary, mule deer remains are found in archaeological sites throughout the plains and mountains but rarely rival bison in numbers. White-tailed deer remains are found less frequently than those of mule deer, but these proportions undoubtedly change in areas of more favorable white-tail habitat farther to both the east and west. Fewer remains are found of elk than of deer, and they are usually in the form of tools and other items manufactured from antlers. However, the amount of elk remains increases during the Late Prehistoric period. Small amounts of black and grizzly bear remains are found, and their contribution to the prehistoric food supply is poorly understood. Small mammals and birds were likely of more importance as a food source than the archaeological record indicates, because their remains have not survived as well as those of larger mammals. All of these species require different procurement strategies.

CHAPTER 8

Weaponry and Tools
Used by the Hunter

THE EXPERIENCED HUNTER AND THE NOVICE

Robert Edgar of Cody, Wyoming, is an accomplished marksman with a handgun. At a distance of 10 meters, he can light any number of matches attached to a board without missing one—a trick effected by having the bullets barely graze the tips of the matches. Before they hit the ground, with five separate shots, he can shatter as many as five small glass bottles or other objects simultaneously thrown high into the air. He shoots the end off a lighted cigarette held in a person's mouth; and given enough encouragement, he performs this feat by sighting his handgun through a mirror held in one hand and shooting over his shoulder with the other. His expertise with a rifle is equally impressive. He has been an avid and accomplished hunter almost from the time he could walk and carry a firearm. Needless to say, any animal that appears within range, moving or stationary, stands almost no chance of escape. He is one of those rare, self-sufficient persons who can take to the woods with a firearm and a knife and simply disappear for as long as he wishes, emerging only when ready and not because of hardship or necessity.

At the other end of the spectrum is the novice hunter whose idea of becoming familiar with a hunting area is to consult a map provided by the local wildlife management unit. To choose a weapon, our novice hunter consults a salesman at the sporting goods store, who acquaints him with a wide selection of charts and tables on bullet and powder weights, cartridge design, and muzzle velocities. The novice emerges fully equipped with a new firearm enhanced with a high-powered telescopic sight. It is highly likely that he was also coerced in purchasing a bottle of female deer scent that, according to its accompanying brochure, can be sprinkled on the trunk of a tree and is almost guaranteed to bring the big buck deer within easy range.

His wisest move would be to have his new firearm and scope sighted in by a specialist, though he could accomplish this himself if he follows the directions carefully. He then spends part of an afternoon at the local firing range using a bench rest with sand bags to hold the weapon steady while he sights it in on targets placed at measured distances. He gains little feeling for the limitations of the weapon, which weight bullet gives the best results, how much to allow for shooting uphill or downhill, how to take into account wind deflection, or what part of the animal to aim for in order to dispatch it while destroying the least amount of meat. Many novice hunters, and even an occasional more experienced one, express surprise that their success on the firing range is not matched by their performance in the field.

Out in the field, the sudden appearance of game usually brings on a rush of adrenaline, commonly referred to as "buck fever"; his reaction is to blast away at the target animal with uncertain results. He may get lucky with a good shot that drops the animal; but he will probably miss, and the animal will disappear over the next rise or into a patch of brush or timber. Should he wound the animal, he lacks the awareness of how an animal reacts differently depending on where it is hit by a bullet. Not all lethal wounds leave a highly visible blood trail that can easily lead him to the animal, so he might fail to make a close enough inspection and thus not notice signs obvious to the more experienced hunter. Without a mentor to guide the novice hunter through this period of learning, he will need some time before he gains any acceptable level of expertise. There is also a strong possibility that his lack of success will be so frustrating that he will abandon any further pursuit of hunting.

In between these two extremes lies a large population of hunters who have acquired a wide variety of expertise in the use of weapons, knowledge of animal behavior, familiarity with hunting territory, and the ability to contend with adverse conditions in the wild. This is true of bow hunters as well as those using firearms. In recent years, interest in using the atlatl and dart has risen and continues to grow. Eventually there may be sanctioned atlatl hunts similar to those undertaken at present with bows and arrows, cap and ball, and flintlock firearms. Whatever the hunter's choice of weapon—whether firearm, bow and arrow, atlatl and dart, thrusting spear, club, or even a simple snare—care and regular practice in its use are critically important determinants of hunting success.

The final and crucial act in a hunting episode, unless the hunt relies on a communal procurement strategy such as a bison jump over a steep precipice, is the delivery by the hunter of a projectile to produce a lethal wound. Nothing is more frustrating to a hunter than to have a well-executed stalk of an animal come to naught because of a poorly made, carelessly maintained, or inexpertly used weapon. Whatever his choice of weapon, neglect of his equipment can ruin a hunting episode and the hunter has only himself to blame. Failure of weaponry because of undetected flaws in raw materials is entirely different. Hidden flaws in the stone and wood of prehistoric weaponry were probably often undetectable until it was too late; in many cases, the result would have been that the animal that was expected to be food for the family disappeared into the distance. Stone projectile points are far more fragile than those of metal. Especially careful attention to the quality of weapon and tool stone must have been crucial to the preparations made by prehistoric hunters.

The dust bowl days of the 1930s caused a loss of vegetation over much of the Great Plains and the lower elevations of the Rocky Mountains. Sandy areas and loose soils were particularly vulnerable to erosion, and the constant wind exposed flaked stone artifacts, as collectors rapidly discovered and exploited. Of particular interest were the Paleoindian (then called "Early Man") projectile points that were beginning to attract the attention of professional archaeologists, but about which relatively little was known at the time (see Renaud 1932; Wormington 1957; Wormington and Forbis 1965). This situation changed rapidly with the discovery and investigation of the Folsom site in New Mexico (Figgins 1927), the Lindenmeier site in northern Colorado (Roberts 1935, 1936), the Finley site in western Wyoming (Howard, Satterthwaite, and Bache 1941), and the Agate Basin site in eastern Wyoming (Roberts 1943). Most significant, however, were the discovery and investigation during the late 1950s and 1960s of the Hell Gap site in southeastern Wyoming (Irwin-Williams et al. 1973). This site produced a stratified, radiocarbon-dated sequence of Paleoindian artifacts that made possible a more systematic study of projectile points and Plains Paleoindian large mammal hunters. The initial aim was mainly to establish a chronology based on point typology; later on, interest expanded into studies of how raw material for the points was procured, how they were manufactured and refurbished, and what their different functional attributes were.

STONE TOOLS AND THEIR USE

In my own conceptual framework, there is a clear distinction between a weapon and a tool: the former brought about the death of an animal and the latter was used in subsequent butchering and processing. Projectile points in animal kill sites are usually assumed to have been the agents of death. However, as methods of use-wear analysis pioneered by Sergei Semenov (1964) gained acceptance, these direct cause-and-effect claims became less straightforward: analyses of some projectile points in some kill sites demonstrate that besides functioning as weapons, the points have undergone post-kill use as tools, probably in butchering and processing the animals. On the other hand, some broken projectile points that were originally of very careful manufacture, fashioned by someone with a definite form (or mental template) in mind, came to show generally inferior quality as the result of being reworked after they were broken, with portions salvaged to be reshaped into, once again, functional projectile points. To cloud the picture further, some broken projectile points were deliberately reworked into tools. A certain amount of butchering was necessary in kill sites involving large animals before the desired parts could be moved to other locations. In some cases, muscles were stripped from carcasses with little if any disarticulation of joints; in other cases, only certain articulated units were removed. Those undertaking these operations may have relied to varying extents on the same or parts of the same projectile points used initially to kill the animals.

Yet it is an oversimplification to claim that one need only to retrieve a projectile point used to kill an animal for it to be suitable for butchering the same animal, especially if, like a bison or elephant, the animal in question is large with thick, tough hide. The binding that holds a stone projectile in a shaft is a compromise between the need to hold the point in place and the need for the point and shaft to pass without hindrance through the hide and into the interior of an animal. The force needed to penetrate a hide with a projectile is applied in a single direction to the base of the point; the forces needed to cut hide and perform other necessary tasks are applied in several directions throughout the process of butchering an animal. In the case of large animals with thick hides, projectile points bound to a foreshaft retrieved from a dead animal require

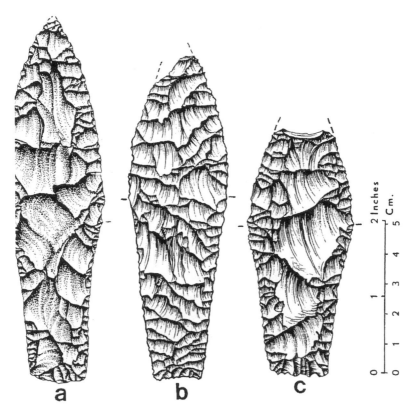

FIGURE 49. Hell Gap–type projectile points from the Casper site. (*a,* from Frison 1991b: 61; *b* and *c,* from Frison 1974: 79.)

some modification to become good butchering tools. In other words, the binding on the projectile during impact is not designed to withstand the forces encountered during butchering. In addition, the sinew binding holding a point in a shaft is usually damaged to some extent when it passes through the animal's hide, as blood and other fluids soften the sinew and stretch and alter the bond between the wooden shaft and the stone point. Heavier bindings can be applied or the projectile point can be mounted in a different kind of handle altogether to provide the leverage needed for butchering. However, if the projectile point is of the right size—such as, for example, large Hell Gap points (figure 49)—it could have been used as a butchering tool without the addition of a handle.

A stone or metal edge that will cut elephant hide is short-lived; the same edge has a slightly longer, but still limited, useful life when cutting bison hide, which is about half as thick. As the edge dulls, the pressure

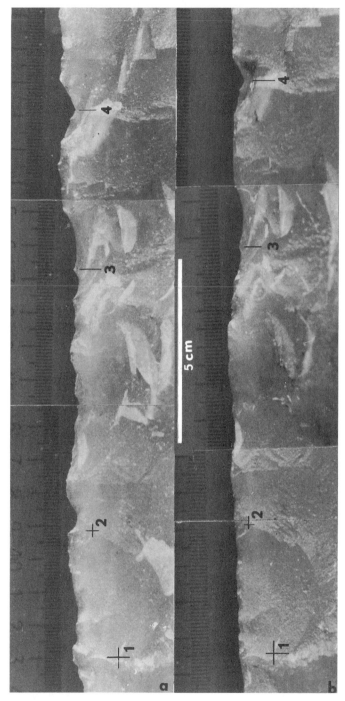

FIGURE 50. Sharp working edge on a 20 cm segment of a
chert biface *(a)* and the same edge dulled by cutting bison hide
(b). The numbers are reference points to identical locations.
(From Frison 1979: 264.)

needed to cut hide increases until a point is reached when the edge simply must be sharpened. Cutting thick hide is the part of butchering that dulls a tool edge most quickly (yet a tool too dull to cut hide can long remain useful in cutting flesh). Most hide retains enough abrasive particulate material on its outer surface to dull either a metal or stone edge quite rapidly. There are visible and easily verifiable differences in the working edge of a stone tool that will cut bison hide and the same edge after it has become dulled and requires sharpening (figure 50).

STONE PROJECTILE POINTS

From my own experience using different types of chipped stone projectiles on large mammals, I have found that some will provide better results than others, though the reasons for the difference are not always readily apparent. Attached to thrusting spears as well as atlatl and dart shafts, some penetrated elephant and bison hide with surprising ease, while the points of others crumbled on impact without penetration or suffered frequent point breakage or damage to the wooden and sinew binding components or both. Any projectile I used that demonstrated good qualities of penetration received special care and was reused again and again for that purpose. Reusing a projectile as a hide-cutting tool, however, rapidly dulls its edges, which require repeated sharpening that soon detracts from its potential as a weapon. In the case of my experiments on African elephants (see chapter 3), I either discarded those projectiles that performed inadequately as weapons or demoted them to the tool category.

The study of prehistoric lithic technology has progressed in positive ways during the past few decades. I spent my younger years in an area where flaked stone projectile points and tools could be picked up just about any time and place the ground was exposed. When I asked about their manufacture, I was told by my grandfather and an uncle that the stone was first heated and that the flakes were removed by carefully applying drops of water. Unable to create anything more than heat-fractured rock in attempts to apply this method, I was then informed that stone flaking was a lost art. One day, I related this story to an old sheepherder who gave me a pitying look and invited me to stop by his sheep wagon, that Spartan prototype of the present-day travel trailer. After rummaging around in a leather saddle bag, he produced a small canvas sack that contained a small river cobble battered on one end; a piece of a broom handle about 20 centimeters long, with a copper nail driven into one end and filed to a dull point; a piece of reddish-colored stone; and several flaked stone projectile points made from the same stone. Using the cobble as a hammer, he drove a flake off the stone and, by repeatedly applying pressure to carefully selected locations on the detached flake with the copper-tipped piece of broom handle, within a few minutes had produced an acceptable projectile point. Later on, I showed it to the proponents of the heat-and-water method, but they refused either to concede that their claims were erroneous or to defend them through actual demonstration. I still have the point the sheepherder made and occasionally use it as a reminder not to accept statements that are based on hearsay or lack satisfactory experimental confirmation. The sheepherder died soon after, so I was never able to learn where and how he learned his flint-knapping skills. I tried to become a sufficiently proficient flint knapper only to better understand the basic principles of lithic technology. I was also well aware that if I acquired too much expertise in this area, I would most likely be accused of making the artifacts I claimed to be finding and bringing in from the hills.

THE BOW AND ARROW

After rising to an above-average level of proficiency in hunting with firearms over a period of years, I was shown a wooden bow and several arrows, found in a Native American burial, and became curious about their usefulness as a hunting weapon. The first bow I made was of willow, because the wood was plentiful in long straight sections free of knots. However, I discarded it after one limb snapped before I was able to pull it to full draw. Willow is strong enough for arrow shafts but lacks the resiliency needed for a bow. At that point, I had saved enough money to buy a cheap bow advertised in the Montgomery Ward catalog, which lasted several months before one limb broke. My choice then was either to attempt to manufacture a better bow or to abandon the entire effort. As I wavered, inspiration to continue came from an unexpected source.

Nick Wilson came by wagon train as a young boy to the Salt Lake area in the mid-1800s. Through a series of extraordinary events, he spent two years as the adopted son of Chief Washakie, at that time the leader of the Wind River Shoshoni, a tribe of historic nomadic hunters in the Montana-Wyoming-Utah area. Wilson's observations during this period provide a unique window through which to view this group of Native American horse nomads as they were acquiring some European trade items but still retained much of their old ways of life. I eagerly pursued Wilson's published accounts of daily life with the Wind River Shoshoni.

FIGURE 51. Juniper tree with a section removed for bow material. (From Frison 1991b: 365. Photo by C. A. Reher.)

He claimed (E. Wilson and Driggs 1919: 107) that most of the Shoshoni used bows made of white cedar (probably juniper, *Juniperus* sp.) with sinew glued to the back. I remembered that the one I had seen earlier, recovered from a Native American burial, had pieces of sinew adhering to its back side, but at that time I could conceive of no explanation for the sinew's purpose. Consequently, I settled for bows made of juniper but without sinew backing—a backing, as I learned later, that markedly improved the performance of wooden bows.

Locating a juniper tree with enough straight, knot-free wood to produce a bow can be extremely time-consuming. A few living juniper trees with sections removed earlier for bow manufacture are still extant, and they clearly demonstrate the strategy involved. Two horizontal cuts were made, one at each end of a section of a standing tree trunk, and the desired part was subsequently removed with wedges (figure 51). I am aware of two living junipers in Wyoming and a dead one in northern Colorado that display the same treatment.

A shrub known as skunk brush *(Rhus trilobata)* was also used in Native American bow manufacture on parts of the plains and in the northern Rocky Mountains, although suitable lengths of usable wood are even more difficult to locate for it than for juniper. Chokecherry *(Prunus de-*

missa) is yet another wood capable of producing a serviceable bow and is more easily found in desirable lengths. Different species of wood, especially ash (*Fraxinus* sp.) and osage orange *(Maclura pomifera),* found toward the eastern parts of the plains, were used extensively in bow manufacture. George Grinnell (1923: 1.173) claimed that for the Cheyenne Indians, "A certain juniper tree (*Juniperus scopulorum,* Sarg.) was regarded as furnishing the best bow wood used in later times," and that "after they had reached the country where it grew, osage orange *(bois d'arc)* was used to some extent, more especially in the southern country where juniper was not always to be had."

Still eager to learn more about the manufacture and use of bows, I happened across another source of information during the 1930s. Robert Cole was an accomplished hunter, blacksmith, and bowman who lived in northern Wyoming in the early and mid–twentieth century. During several decades of making and using several bows, Cole settled on yew wood (*Taxus* sp.), which grows in northwestern North America, as better than any wood native to the Wyoming area. In addition, it yielded satisfactory results without a sinew backing. However, yew wood was difficult and expensive to obtain; and as transverse breaks would develop in the limbs of his bows, Cole was often forced (mostly for economic reasons) to salvage the central parts of two bows and splice them together to make a single one.

Cole was a large man and reportedly used a bow with a pull of about 100 pounds, practicing almost every day. He was a meticulous craftsman who turned his wooden arrow shafts on a lathe; he used a metal template to form his triple-bladed steel points, honed to razor sharpness, and another template to apply fletching to his arrow shafts. At the Labor Day rodeo in the town of Thermopolis, Wyoming, in the summer of 1939, he stood at one side of the rodeo arena and drove an arrow through a mature male bison. The animal was butchered and prepared for a free barbecue for everyone attending the rodeo celebration (see the *Thermopolis Independent Record,* September 7, 1939). The animal chosen was the oldest male from the herd and well beyond further use in breeding cows. Consequently, the meat was tough, tasteless, and a poor substitute for the flesh of a young animal, and the barbecue resulted in an understandable general dislike for bison meat in the Thermopolis area for several years thereafter.

I am sure that Cole regarded me as a pest for continually badgering him to teach me the art of the manufacture and use of a bow and arrow. I remember him going into a prairie dog town one day and literally nailing one animal after another to the ground. Several people testified that he once killed a large bull elk at a measured distance of 175 steps. My own expertise with a bow and arrow went no further than hunting rabbits, prairie dogs, and marmots. Cole took one look at my archery equipment and stressed that if I wanted to gain any real success in hunting large game, I would have to acquire a better bow, take more care in the manufacture and assembly of arrow shafts, and, above all, practice every day.

William Maycock, whose account of pronghorn behavior has already been noted in chapter 5, was also accomplished with a bow and annually guided numerous bow hunters. He claimed that a successful bow hunt of antelope required that the shaft be put in the chest cavity; otherwise, the result would be another wounded animal requiring time and effort to locate. He also stressed the necessity of continual practice and, for the best outcome, the need to use a bow that has all the power the hunter can pull (Maycock 1980: 79–80).

Nick Wilson claimed that the best bows the Shoshoni had were made of mountain sheep horn backed with sinew (E. Wilson and Driggs 1919: 107). One, reported to have been recovered from a cave in the Gros Ventre Mountains in northwest Wyoming during the first decade of the twentieth century, was preserved down through the years (figure 52), and most of the details of its manufacture can be reconstructed. It was made from strips taken from two ram's horns, probably a pair, joined at the center with a peg, also of horn, that was 2.3 millimeters in diameter. The piece of material over the center, probably a thick section of horn, that held the two sections rigid is missing; from the change in coloration where it was attached, this piece was about 13 centimeters long. The bow's total length is 83 centimeters, with the greatest width of 4.8 cm at its center. A continuous sinew backing, 5.5 millimeters thick at the center of the bow, was glued to the horn, which is 10.5 millimeters thick at the same spot; the total thickness is thus 16 millimeters. The bow is a reflexed type (Frison 1980) and is presently on display at the Museum of the Mountain Man in Pinedale, Wyoming, close to where it was discovered. It is one of few authentic mountain sheep horn bows presently known to exist.

According to Wilson, the sheep horns, which needed to be those of a mature ram to obtain the necessary length, were first thrown into a hot spring (which are common in Shoshonean territory) until they became pliable and strips the length of the horn could be removed. The sinew was applied to the back with "some kind of balsam gum" glue (E. Wilson and Driggs 1919: 107). According to another account, the horn was

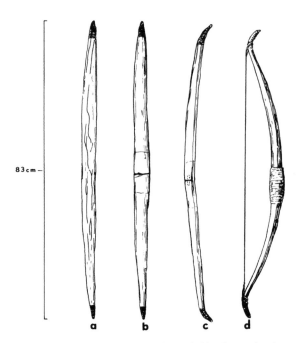

FIGURE 52. Sinew side *(a)*, horn belly *(b)*, and side
view *(c)* of a reflexed, sinew-backed mountain sheep
horn bow. In *d*, the same bow is sketched with hand-
grip added and in braced position. (From Frison
1980: 75.)

softened by being slowly heated over a bed of coals and each piece of
horn was also beveled, overlapped, and glued together at the center of
the bow. The glue to attach the sinew was made by boiling "shavings
from the hoof and small bits of thick neck-skin or back skin," until a
"thick scum formed and was skimmed off" (Dominick 1964: 155–56).
A thick piece of horn about 5 inches (12.5 cm) long was placed over the
junction of the two sections and wrapped with wet rawhide, which made
a firm joint when it dried. It was claimed that "it took two months for
a skilled specialist to turn out such a bow, and other Shoshoni people
and even people of other tribes traded eagerly for them" (Dominick
1964: 156).

    A bow on display in the Pioneer Museum in Lander, Wyoming, was
made from two staves taken from the main beams of a pair of elk antlers
(figure 53). The back of the bow is sinew, about 2 millimeters thick, applied to the outward side of the antler. The inside, or belly, is smoothed,

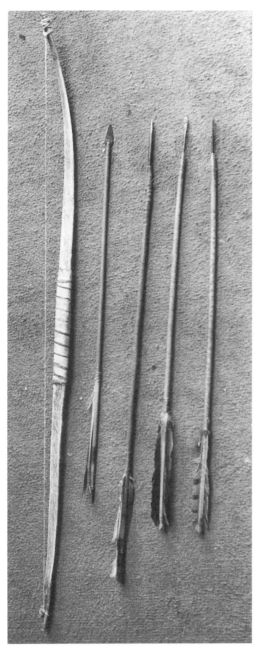

FIGURE 53. Sinew-backed elk antler stave
bow and arrows. (Photo by author. Courtesy
of the Museum of the American West, Lander,
Wyoming.)

with all of the porous part removed. Its total length is 109 centimeters and its maximum width is 27.4 millimeters. The two staves are beveled and joined at the center. A 20.5-centimeter-long piece, probably antler, covers the center splice and is held in place with a heavy sinew wrapping. Thin hide, probably pronghorn, covers the entire central area. One limb is snapped a short distance from the end and the bow string is made of sinew. Arrows included with the bow have two types of tips: some are pieces of iron roughly hammered flat into the shape of a projectile point, and the others are pieces of round metal rod pointed on one end.

Perhaps no two individuals have invested more time and effort into studying, making, and using Native American bows than Reginald and Gladys Laubin. They say osage orange and yew are the best bow woods, but also mention ash, chokecherry, and cedar as acceptable. Both Laubins are highly accomplished technologists. They have provided detailed accounts of the entire process, from cutting the strips from the horns to adding the sinew backing, in their manufacture of mountain sheep horn bows (Laubin and Laubin 1980: 73–88). Although the Laubins performed exhaustive experiments with elk antlers, they were never able to manufacture a bow from them that they considered satisfactory (Laubin and Laubin 1980: 73–103). One observation that emerges repeatedly from their experiments with all types of sinew backing is that a period of at least several months is required for proper curing, during which time the bow must be properly hung (see also Bergman and McEwen 1997).

Grinnell described the manufacture of elk antler bows among the Cheyenne Indians in some detail, but at the end he comes to much the same conclusions as the Laubins: "such bows were fine to look at, but they were more for show than looks. They did not last long, but were likely to break when they became old" (Grinnell 1923: 1.174). He also mentions bows made of pieces of buffalo horn glued together.

Clark Wissler briefly discusses Blackfoot Indian weapons but cautions the reader that they had abandoned the use of bows for so long "that good, practical specimens are scarcely to be found and accurate knowledge concerning them difficult to secure" (Wissler 1910: 155). He illustrates a sinew-backed hunting bow (1910: 156) and mentions smaller wooden ones used for games. Those persons seeking detailed information on North American Indian bows, arrows, and quivers should consult Otis Mason's treatise (1893) on the subject.

Other than the mountain sheep horn bow, the Plains Indian bows I am aware of are adequate but not among the better performers. William

Laughlin's comments on various bow and arrow users worldwide reflect much of my own thinking on the subject.

> The bow and arrow is in common use among many hunters—Pygmies, Bushmen, Eskimos, various American Indians, Andamanese, Chukchee, to name only enough to illustrate considerable diversity in the technology and use; however, most observers agree that these hunters are mediocre or indifferent as archers. They hunt effectively with their equipment, but, they compensate for lack of accuracy at appreciable distances, perhaps more than twenty or thirty yards, by spending their time getting closer to the animal. In brief, these hunters clearly spend more time and attention in utilization of their knowledge of animal behavior than in improvement of their equipment or of its use. (Laughlin 1968: 306)

He emphasizes these ideas further:

> Children were taught to close the distance between themselves and their quarry by sophisticated stalking methods that depended more upon comprehensive observation, detailed ethological knowledge and an equally detailed system of interpretation and action than upon the improvement of their equipment and the addition of ten or twenty yards to its effective range. In fact, one may pass from this generalization to another and suggest that the very slow improvement in technology, clubs, spears, throwing boards, bows and arrows, as indicated by the archaeological record, was contingent upon success in learning animal behavior. It was easier or more effective to instruct children in ethology, to take up the slack by minimizing their distance from the animal prey, than to invest heavily in equipment improvement. (Laughlin 1968: 306)

In a similar vein, I contend that it was possible to improve one's ability to get closer to the animal, but the raw materials and technology for manufacture that were available limited improvements in weaponry. These same caveats apply to the atlatl and dart as well as to the bow and arrow.

THE ATLATL AND DART

Numerous people, avocational and professional archaeologists alike, questioned that an atlatl I recovered from a dry cave in the Big Horn Mountains of northern Wyoming (figure 54) was well enough designed and large enough to be effective on animals the size of deer and pronghorn, let alone bison. However, the one nearly complete specimen and the parts of three others were of similar dimensions, suggesting a commonly used weaponry type rather than an aberrant one. The wood used in manufacture was skunk brush, mentioned earlier as used in Native

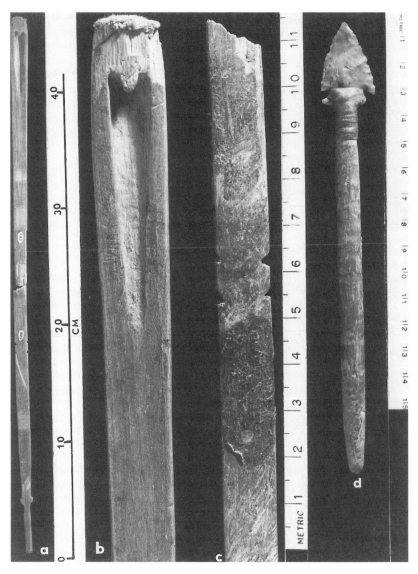

FIGURE 54. Atlatl *(a–c)* and foreshaft *(d)* with projectile point from Spring Creek Cave. (From Frison 1978: 60.)

American bow manufacture—an extremely tough, resilient wood though difficult to find in usable lengths and diameters. Closely copied replicas of the original atlatl, the associated shaft material, and stone projectile points proved adequate for killing jackrabbits, cottontails, and marmots. I never tried it on deer or antelope, but I am convinced that with sufficient practice to improve accuracy, it would prove reasonably effective for those animals. I had no further experience with atlatl and dart until the opportunity to experiment with Clovis projectile points on African elephants required heavier weaponry components.

After committing myself to the testing of Clovis points on African elephants, I realized it would be prudent to revive my earlier skills, however limited, with atlatl and dart. I was convinced that the ones found and replicated earlier would not produce the same results on elephants as on jackrabbits, so I made three atlatls, all of skunk brush, and several dart shafts of willow—all of larger dimensions than my earlier experiments. Through repeated trial and error, I found that one atlatl produced the best results, so I discarded the other two. The willow shafts performed well in penetrating bales of straw but eventually proved inadequate to the much more difficult task of penetrating the hides of mature African elephants. Consequently, I changed to shafts made of chokecherry, more resilient than willow though more difficult to obtain in desirable lengths. However, chokecherry has one major disadvantage: it never cures to the point that it is no longer susceptible to warping and twisting. Unable to locate another wood suitable for shafts, I spent long hours maintaining my weaponry in addition to attempting to acquire enough experience to develop some confidence in its use.

In earlier chapters, I have already discussed the superior quality of Clovis projectile points and the problems encountered in using and maintaining atlatl and darts when pursuing African elephants, as well as the use of other projectile points in bison procurement. However, another discovery in the use of an atlatl and dart came about by accident during the experiments on African elephants. I used two mainshafts in the experiments: one was 202 millimeters long, with a maximum diameter of 23.5 millimeters and a weight of 365 grams; the other was 198 millimeters in length, with a maximum diameter of 22.5 millimeters and a weight of 358 grams. To avoid problems while carrying them on airlines, I made them in two sections. However, the joints would withstand only short use before breaking under the stress of penetrating elephant hide. Fortunately I carried along an extra section of shaft, 25 centimeters long, that I split into thin strips. I used high-quality elephant sinew, which was readily available in substantial lengths, to bind three of the strips to the outside of the smaller of the two shafts, ensuring that they were 120 degrees apart and extended for equal distances on both sides of the joint. This reinforcement solved the breakage problem and, in addition, increased the shaft weight from 365 to 420 grams. The added weight improved performance, but further increases made the weapon difficult to control and required too much added thrust. Continued experiments convinced me that to create the optimum hunting equipment it was necessary to take into account a number of factors, including the user's body size, weight, and arm length; the atlatl and mainshaft length and weight; and the foreshaft and projectile weight. A long period of trial and error is needed to bring these several variables into perfect harmony; once that harmony is achieved, the resulting weaponry will become one of the most personal of human possessions. I have tried several different atlatls and darts made by others and have not been unable to achieve results nearly equal to those obtained using the ones with which I had become intimately familiar.

The joint formed by a tapered foreshaft inserted into a matching conical hole in a mainshaft functions adequately, as long as care is taken to match the angles and contact surfaces of both elements very closely: a loosely fitting joint is an invitation to broken equipment, lost animals, and time-consuming repairs. A heavy sinew wrapping on the distal end of a mainshaft prevents its being split by the foreshaft on impact (see figure 7). Light sinew wrappings applied also to the proximal or cup end of mainshafts prevent splitting by the atlatl spur. Many tapered foreshaft ends recovered from dry caves have a light spiral rasping (figure 55). Experiments with replicas of these foreshafts demonstrate that the hunter needs only to moisten the tapered end lightly with his tongue, insert it into a mainshaft, and apply a quick twist in the direction of the rasping to seat it firmly. This kind of joint withstands impact but allows easy separation afterward; the hunter can thus retrieve the mainshaft and insert another foreshaft, and thereby be relieved of the added burden of carrying extra mainshafts.

The nearly complete atlatl that I recovered had shallow notches on both edges of the shaft near the proximal end and patches of unidentified dried glue on the upper side adjacent to the notches, suggesting that some object, probably an atlatl weight, had originally been secured there. In addition, one complete and one broken carved steatite artifact from the same cave site component appear to be atlatl weights. One side of the broken

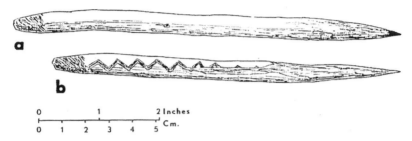

FIGURE 55. Wooden foreshafts with sharpened distal ends and spirally rasped proximal ends from a Late Archaic level in Wedding of the Waters Cave in northern Wyoming (a, b). Note a decorative or possible ownership mark on b. (From Frison 1962: 251.)

one has a longitudinal concavity that fits closely the curve of the atlatl's back; it also has the same residue in the concavity as that found on the atlatl (figure 54c). There are two general explanations of atlatl weights: one is that they represent some sort of repository of supernatural power passed from one generation to another, and the other is that the added weight somehow improves the performance of the atlatl. I have tried several different weights in my atlatl experiments without detecting any improvement when they were added. However, these experiments lacked the controls necessary either to be convincing or to produce acceptable results.

Further experiments soon confirmed that it requires a strong application of force from the atlatl spur to the proximal end of a dart shaft to achieve sufficient velocity to penetrate a mature bison's hide, with even more force required in the case of an elephant. The rapid body movement needed to accomplish this maneuver can cause an animal watching the hunter to quickly change its position. The time that elapses between the moment when the forward movement of the arm begins and when the projectile reaches its intended destination may be long enough to allow many animals to react swiftly enough to avoid the projectile. An added consideration is that the direction in which the animal will move is not always predictable. This claim is easily tested and proven: a deer that suddenly appears along the side of the highway may react to a moving vehicle in a number of different ways. It may continue its forward progress across the highway, or it may stop and let the vehicle go by, or it may reverse direction and retrace its steps, or it may stop and then suddenly try to cross in front of the vehicle, or it may move ahead a step or two and then reverse direction. Similar reactions occur when the sudden appearance of a human hunter startles an animal. Given this extreme uncertainty about how an animal will respond to the sudden movement by a person casting a projectile with an atlatl, my personal preference is to have an animal committed to a direction of movement before the throw: the hunter then need only allow enough lead for the projectile and target to meet at the appropriate spot. However, this strategy works only if the animal is moving at close to a right angle to the hunter. One should always avoid shooting at an animal moving directly away. These kinds of problems are less severe with bow and arrow, both because the body movement required in pulling and releasing a bow is less obvious and because the greater velocity of the projectile allows the animal less time to react. Two hunters experienced in operating together often have an advantage. While the animal is watching one hunter, the other hunter may be in a position to take the shot.

ITEMS OF IVORY, BONE, AND ANTLER

Certain objects made of ivory, bone, and antler are often believed to be projectile points, though their true function in some cases may be open to question. For example, the Anzick Clovis cache produced cylindrical-shaped objects of bone with cross-hachured, single-beveled ends (figure 56b); one had one end conically shaped (figure 56c), much like the proximal ends of known wooden foreshafts of Late Archaic age (figure 54d). One proposal is that these were used as foreshafts for Clovis points (Lahren and Bonnichsen 1974). However, bone objects of similar size to but different morphology from the Richey Clovis cache (Gramley 1993) demonstrate unique wear patterns, leading to speculation that they might have been incorporated to form sled runners (a suggestion that to me seems doubtful). A cylindrical object of ivory, presumably mammoth, with a single-beveled, hachured end and a missing distal end (figure 56a) was recovered at the Sheaman Clovis site (Frison and Craig 1982: 157) at the Agate Basin site locality in eastern Wyoming. What remains of the distal end suggests that it originally had a long, sharp point similar to those made of ivory recovered in Florida (see Gramley 1990: 40a).

The bison and pronghorn bone bed in a Folsom level at the Agate Basin site (see figure 31) produced objects made of elk antler strips that appear to be projectile points (Frison and Zeimens 1980). One is in two pieces

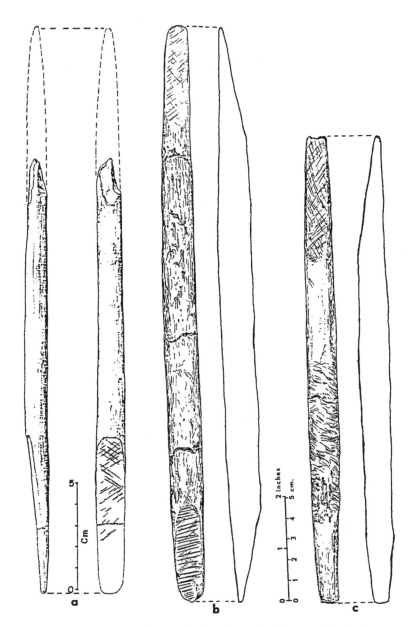

FIGURE 56. Cylindrical ivory shaft *(a)* from the Sheaman Clovis site (Frison 1982f) and bone shafts from the Anzick site *(b, c)*. (From Frison 1991b: 43; Lahren & Bonnichsen 1974: 148.)

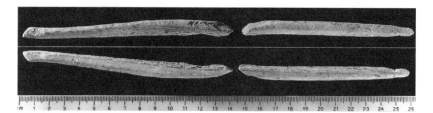

FIGURE 57. Elk antler point from a Folsom component at the Agate Basin site. (From Frison & Zeimens 1980: 233.)

and, although a part of the center is missing, the longitudinal marks on both pieces where it was cut from the beam of an elk antler match, leaving no doubt that both pieces are part of the same specimen (figure 57). The distal end is round in transverse cross section and tapers to a sharp point, while the proximal end is rectangular in cross section. A short triangular or trapezoidal section from its center is missing, and experiments with replicas to penetrate freshly killed elk and domestic cow carcasses confirmed this to be a common breakage pattern (figure 58). Such breakage occurs when a point encounters bone, causing a sudden change in direction.

Two broken bone artifacts from the Lindenmeier Folsom site are probably examples of the same artifact type as the ones from the Agate Basin site (Wilmsen and Roberts 1978: 131). In fact, there is a remarkable similarity in worked bone artifacts from both the Agate Basin and Lindenmeier site Folsom components, which have almost identical radiocarbon dates (see Frison 1991b: 25). A probable bone projectile point was recovered at the Late Prehistoric Vore Buffalo Jump. It was ground to its present shape, 80 millimeters long with a square cross section 8 millimeters wide. It tapers to a sharp distal end and has a cone-shaped base 17 millimeters long (Reher and Frison 1980: 27–28).

Bone or antler points can penetrate thick hide, but they are unable to cut the hole needed for the entry of the foreshaft. Using an atlatl, I had difficulty getting an experimental bone point through the hide of a mature male bison: in this case the projectile penetrated for a distance of about 6 centimeters. The circumference of the hole was still increasing when the point stopped; and the hide gripped it so tightly that it snapped when I tried to pull it out of the carcass. The remains of pronghorn, animals with hide much thinner than that of bison, were recovered in the

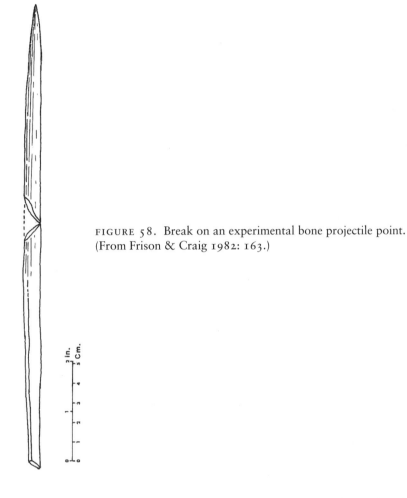

FIGURE 58. Break on an experimental bone projectile point. (From Frison & Craig 1982: 163.)

Folsom-age bison bone bed at the Agate Basin site and might explain the presence of this type of point. To speculate further, these points might have been used in killing smaller mammals, such as scavengers attracted to the site. But the relative rarity of bone and antler points in animal kills suggests they were not preferred, at least for use on large mammals such as bison.

Heidi Knecht, after extensive experiments with copies of early Upper Paleolithic bone and antler points, concluded that antler points are easier to manufacture than bone points and suffer less impact damage than bone points; she also suggests that because of breakage patterns, they can endure more "rejuvenation episodes" than bone points. However, she qualifies the latter claim with a need for better verification (Knecht 1997: 206). I am well aware that controlled experiments with copies of North American bone and antler projectile points are needed to produce conclusive evidence.

WOODEN AND METAL PROJECTILE POINTS

Also recovered in the Big Horn Mountain dry caves were wooden fore-shafts with sharpened distal ends (see figure 55), one with an enlarged distal end, possibly a blunt point, and several with battered distal ends (see Frison 1965: 87). These may have been used to obtain small animals and birds. I was able to reproduce closely the battering on the ends of experimental foreshafts by using them for practice. Expertise with the atlatl and dart comes only with continued and intense practice, and those practicing must have been reluctant to unnecessarily waste the points designed and prepared for serious hunting.

With the introduction of European trade goods to the Native American tribes, metal projectile points became very popular. Some were also made by the Native Americans themselves employing a crude form of metalworking on material salvaged from sources such as the iron straps used to reinforce wagon parts. Brass, probably less readily available, was used less frequently. One brass point, made from what appears to be a brass cartridge case hammered flat, was recovered from the camp site at the Big Goose Creek site in northern Wyoming (Frison, Wilson, and Walker 1978: fig. 15L). These metal points withstood the forces of impact better than those of stone. The one mentioned in chapter 2 as sticking in a bison horn's sheath was bent at nearly a right angle but was easily restored to its original shape. Both Native American and European metal trade points were common surface finds several decades ago, but most have long since rusted away.

In his discussion of the Cheyenne Indians' early contacts with whites, Grinnell relates the story of a starving man who wandered into a Cheyenne camp, where he regained his health. Later he returned to them with European goods, including metal: "With pieces of iron that he brought, he made arrowpoints which they at once saw were better than theirs of stone or bone, so that all who could procure it used the iron for arrowpoints. With these arrows they could kill animals much more easily than with those made of stone. They, therefore, threw away their stone points" (Grinnell 1923: 1.34). How much of this account is truth and how much fiction is difficult, if not impossible, to determine, but I am convinced of the superiority of a metal projectile point over one of stone.

RED OCHRE, ROCK ART, AND WEAPONS

Red ochre appears with the Clovis cache assemblages and in many human burials, strongly suggesting a ritual context. The red ochre covering fragments of two juvenile human crania recovered with the Anzick Clovis cache implies some ritual connection. Red ochre is commonly found distributed throughout Paleoindian cultural levels. It has many uses: it is a preservative needing only a carrier such as animal fat to provide a protective covering for wood. The Archaic-age atlatl and dart parts described in chapter 2 were heavily coated with red ochre. It is also an abrasive (commonly known as *jewelers' rouge*) that, applied to a strip of tanned hide or cloth, can be used to polish stone. Gene Titmus, an able flint knapper who became deeply involved in Clovis lithic technology (see Titmus 2002), once demonstrated to me that he could use ochre on a strip of hide to dull the blade edges near the base of Clovis points, thereby producing the results found on actual Clovis points. However, no unequivocal evidence has been found to confirm that this method was used during Clovis times.

One source of high-quality red ochre that was mined during Paleoindian times is located in eastern Wyoming a short distance from the Hell Gap site. In the mine tailings, along with a large assemblage of simple flake, biface, and bone tools, are complete, reworked, and broken projectile points characteristic of all the Paleoindian cultural complexes known to the area (Tankersley et al. 1995; Stafford et al. 2003). The past use of red ochre is complex and far from being satisfactorily understood, but it does appear to have played a significant role in prehistoric hunting equipment, perhaps used for ritualistic purposes, as a preservative, as an abrasive, or possibly in all three ways.

Some of the pecked, painted, and incised figures on rock faces appear to have a strong connection with hunting. However, these representations nearly always relate to the Late Prehistoric and Historic periods. Animal figures are regularly shown with arrows penetrating various parts of their bodies (figure 59) and were likely closely connected with hunting magic (see Francis 1991; Sundstrom 1984). In most of these figures the fletching is exaggerated in size and of a different shape from that known to appear on actual specimens. I am not aware of rock art figures found in the plains or Rocky Mountain areas that unequivocally depict

FIGURE 59. Petroglyph panel with animal figures penetrated by arrows.
(From Frison 1978: 417. Photo by C. R. Swaim.)

the use of the atlatl and dart and thus might indicate its use extending
back into the Archaic and earlier periods.

OTHER IDEAS ON WEAPONRY

Peter Bleed (1986) has developed two concepts regarding the design of
hunting weaponry: one was maintainability and the other reliability. Margaret Nelson (1997: 377) adds to these the concepts of durability and
use-efficiency. I would add portability; a hunting episode may terminate
a few minutes away from camp or it may last well into evening dark-
ness. On a long hunt the needed equipment, which must always be kept
available and ready for use, can become extremely burdensome. Aes-
thetics is also a consideration: most hunters take pride in the appearance
of their weapons and, besides keeping them in top functional condition,
tend to embellish them with nonfunctional decoration. These ideas
about the range of hunter behavior are gleaned from my reflections on
the attitudes of the hundreds of hunters I have associated with for more
than seventy years; I believe that the hunters of earlier ages were not so
very different.

    A cautionary note is appropriate here: I have never considered using a
bow for any purpose other than to propel an arrow. However, this is not

universally true: Russell Greaves (1997: 301–7) portrays Pume hunters in Venezuela using bows to probe for, dig out, and club rodents. I would regard this kind of harsh use as damaging to any bow with which I am familiar, but it carries a strong message to not close one's mind to considering alternative uses of weaponry.

The weaponry used for more than 11,000 years of hunting on the plains, in the Rocky Mountains, and in parts of adjacent areas varies to some extent but still conforms to a single design and purpose: a stone, metal, bone, or antler projectile with a sharp tip is propelled with enough force to penetrate and deliver a mortal wound. A hunter using weaponry such as the atlatl and dart and the bow and arrow required extensive practice to gain the proficiency needed to provision a family. Additionally, the manufacture and use of weaponry appear to be embedded in many prehistoric cultural institutions in ways often difficult, if not impossible, to extract from the archaeological record.

CHAPTER 9

Concluding Thoughts

After a lifetime of hunting, guiding, trapping predators, and observing wild animals in their natural habitat, I find several thoughts about human and animal relationships coming to mind. For one thing, there is a difference in the way in which human and animal predators affect their prey. Four-legged predators attack animal populations from the bottom, first seeking out the old, the weak, the very young, and the crippled; human predators attack from the top, choosing if they can the individual animals in the best condition that promise to provide the most and best-tasting meat. If there is a good balance between predators and the prey populations, the former do little damage to the latter. However, too many animal predators can throw the system out of balance, with serious consequences for both prey and predators. A good example is the recent increase in mountain lions, which have seriously depleted both mule deer and mountain sheep populations in parts of the western United States. In some locations, they have so reduced the deer population that, having destroyed their food source, the mountain lions have been forced to move on.

Human predators, targeting the strong, remove animals needed to continue the health and growth of animal populations. Depending on the intensity and effectiveness of their hunting, and unless properly regulated, human predators as well as the four-legged ones can undermine the viability of their own food supply. As one example, we need only to recall the near loss of bison and other animals on the plains and in the mountains because of overexploitation. In this sense, human predators also function as a continual drain on the animal populations that constitute their subsistence base. With both animal and human predators the balance is at best precarious, and easily upset.

The life of a hunter can be conceptualized as a complex of continually changing emotions. A hunting episode begins with expectations of finding the animal of choice in a location where past experience predicts it should be. If all goes well, there is a feeling of euphoria at success, immediately followed by a fleeting feeling of remorse at taking the life of a prime animal. Following this, there is the drudgery of field dressing and getting the animal back to camp. The return home is accompanied by the expectation of approval by family members. From my own past experience, I know that the hunter returning to camp with an inferior animal will be met with disapproval from family and derision from peers, both (again from my own experience) strong motivators to exert maximum effort to bring home a prime animal.

Expectations of success often go awry as a result of the unanticipated: the wind changes, predators are lurking nearby, birdcalls alert the animals to the hunter's approach, or the hunter carelessly snaps off a branch. All of these can and often do disrupt a planned hunt; the hunter's only choice is to initiate another attempt, with a mixture of frustration and determination, because returning to camp after an unsuccessful hunt can elicit the same reactions as bringing back an inferior animal. In his observations on the Nunamiut, Lewis Binford (1978: 181) notes that "if a young hunter kills a very 'skinny' bull or cow or even a calf, he may be embarrassed to return his 'prize' to the village, since he can expect much teasing." In any hunting context, these are strong forces that bring out one's best performance; long experience has taught me that being perceived by others as anything less than a competent hunter can be devastating.

Every hunter occasionally experiences periods of varying duration without any success. While such failure can bring on fear and doubt in one's ability, it usually builds up determination and thus leads to greater expression of one's hunting ability. It is a great relief when the period of failure finally ends and, once again, one's family and peers offer the approval and respect that come only with success. In the case of the prehistoric hunter, continual failure could lead to concerns that something was wrong in the supernatural realm. The shaman could begin to probe the hunter's behavior, possibly looking for signs of neglect to observe

proper rituals in the treatment of animal spirits. There are many avenues
to explore in documenting the different societal influences that affect
hunters and hunting (see, e.g., Hawkes 1990).

Informed speculation on the effects of human predation on animal populations is possible in the case of the bison, the highly favored prehistoric food source on the plains. Alexander Henry (Henry and Thompson 1897: 1.444) claimed a traveler's daily allowance of fresh bison meat
was 8 pounds (3.6 kg) or 1.5 pounds (0.68 kg) of pemmican. David
Thompson (1962: 313) declared a century earlier that "even the glut-
tonous french canadian that devours eight pounds of fresh meat every
day is contented with one and a half pound [of pemmican] pr day." Isaac
Cowie (1913: 214–216) boosts the man's daily ration of fresh meat to
12 pounds (5.4 kg), or 3 pounds (1.4 kg) of pemmican, putting a woman's
ration at half that and a child's ration at half that of a woman. Theodore
White (1953b: 163) arrived at different numbers: "A conservative estimate for the meat requirement is taken at one-half pound of lean meat,
or equivalent, per person per day. For rapidly growing children, people
doing hard physical labor, or under exposed weather conditions, the requirements are much higher."

These figures allow some leeway in calculating the daily meat requirements for prehistoric Plains hunters. Following White's thinking, the person pursuing animals and carrying meat back to camp burns many more
calories than the same person in camp resting, taking care of weapons,
and performing other camp chores. From my own experiences hunting
and tending livestock for extended periods away from the ranch headquarters, I believe White's estimate to be very low; 1.5 kilograms (3.3 lbs.)
of fresh meat as an average per person per day requirement is more realistic, though still on the conservative side.

The mature female bison I butchered with help from another person
as described in chapter 4 produced 167 kilograms (368 lbs.) of meat,
which is probably close to the animal's average weight year-round. If the
daily ration per person is 1.5 kilograms, this animal might feed a family
of five for a month (or longer, if small children were in the group). Assuming a band-level social group of five families operating together as a
unit and living exclusively on bison, they would need to kill about five
or six bison a month, or between sixty and seventy animals a year. To
me, this implies a high level of hunting activity. The best strategy would
have been to kill one or two animals at a time and in such a way as to
cause as little disturbance as possible to the remaining animals. These requirements test the hunters' skills and could determine the group's ac-

tivities in the immediate future. For example, a poorly executed hunting episode could cause the animals to be more wary of humans; especially in the case of bison, they might even temporarily abandon an area, forcing the entire hunting group to move in response.

The communal bison or other animal kill had to be a special occasion that served functions besides meat procurement. As several bands came together to collect the number of hunters necessary, it would have been the ideal time for other societal activities. Certain conditions would have had to be met before a communal kill was possible. A sufficient number of animals would have to be available; the location of the trap would have to allow for animals to be driven to it; and, depending on the kind of trapping intended, the proper geologic feature or the trees needed to build a structure would have to be present. It is highly improbable that a five-family band of hunters contained the personnel required to build a trapping complex such as, for example, the Ruby site bison corral and religious structure (see map 4), and then to gather and maneuver a herd of bison into it. Thus two and even three or more bands needed to join forces to carry out a successful communal kill. If we assume an aggregation of three bands, then the number of bison killed in an area during a year would increase to between 180 and 200. Depending on how many of those killed were breeding females, human hunting could have seriously affected the resident bison population's ability to produce enough replacements to continue its existence. Furthermore, such calculations ignore the presence of other predators such as wolves that also partially depended on the bison for their survival.

If the bison were not available or a change in diet was desired, a hunter might temporarily switch to deer or pronghorn. One such animal (if we keep the daily per person requirement at 1.5 kg of meat) would feed the family of five for only four to six days. In this case, I believe, hunting would have to be nearly continual. Allow a day or two to obtain an animal and bring it to camp, another day to refurbish weapons and tools—and then it is time to reconnoiter the territory and contemplate the next hunting episode. The level of activity required would be a true test of the hunter's ability.

An observer of a hunting episode such as a prehistoric communal bison, pronghorn, or mountain sheep kill would have seen the proverbial tip of the iceberg—only a small portion of the entire process. In the eyes of the uninformed, a successful bison drive might have appeared as an event carefully planned and rehearsed beforehand, from start to finish. In reality, however, it involved split-second decisions and immediate reactions, as the animals responded in a range of ways as they attempted to evade the human predators. The hunters are aware of these possible responses but cannot predict their order or magnitude. These are the critical moments, when the hunters' responses to the animals' reactions determine the difference between success and failure. To successfully match wits with wild animals with the intent to kill them requires a thorough knowledge of the hunting territory and the behavioral patterns of the species residing within it.

Ideally, this knowledge is acquired over a long period of time, beginning in early childhood. William Laughlin (1968), whose area of expertise was in the Arctic, emphasized this point many decades ago. One major difference in our positions is that the Arctic provided him the opportunity of observing ongoing Eskimo hunting strategies with weapons and tools whose use went back to prehistoric times. Lacking the same opportunities in my area of interest, I have attempted to use the same kinds of artifacts recovered archaeologically to reestablish past paleo-ecological conditions as nearly as possible, and to rely on present-day animal behavior operating within ecosystems similar to those of the past to interpret past procurement strategies. The flip side of this approach is a conviction that if animal behavior does not appear commensurate with a procurement strategy as indicated in its archaeological context, it is time to step back and rethink the data. The key element is becoming familiar enough with animal behavior to make possible reliable interpretations.

Pursuing this line of thinking, if we begin our analysis of an archaeological bone bed at the time the products of the hunt entered the system, and the events leading up to that moment are ignored, a significant amount of information about the daily activities of the study group will be ignored and possibly lost. There are enough animal procurement sites that are well enough preserved to test the strategies involved against the observed behavior of the prey species involved. It is then up to the investigator to explore all possible avenues of analysis and interpretation, including an understanding of the learned behavior needed to procure the animals.

In addition to the ritual and supernatural activities involved, the size and complexity of bison, pronghorn, and mountain sheep procurement complexes imply some form of authority or social control to initiate and oversee construction of the facilities and to coordinate the activities of gathering and driving animals. Though all of these activities are interwoven into the fabric of the society, they are difficult, if not impossible, to extract from the archaeological record. Yet specific instances—the preservation of parts of a bison corral and an associated religious structure, a shaman's structure incorporated into the drive line of a sheep trap, a circle of bison skulls in a bone bed, deer skulls with antlers attached set in a pattern around another skull placed on top of a rock cairn, and the like—provide rare glimpses of ritual observances in the past and remind us to allow for their presence.

As archaeologists, we do a great deal of postulating on the nature of social groups that were involved in prehistoric hunting, particularly about the late Pleistocene–early Holocene hunters who dealt with now-extinct animals. Edwin Wilmsen (1974), by using the Lindenmeier Folsom site (Wilmsen and Roberts 1978), ideas from human social organization, and data from ethnographic studies, attempted a hypothetical reconstruction of a late Pleistocene hunting group. Even though Wilmsen's effort is now more than a quarter century old, I believe it has considerable merit and is an approach that could still be used to advantage as we attempt to understand prehistoric hunting groups.

There is a cautionary tale for those unaware of the rapidity and intensity of climatic changes that undoubtedly affected prehistoric hunting. As mentioned in chapter 1, extreme weather conditions and the aftereffects of the winter of 1886–87 eliminated many of the range cattle on the Northern Great Plains (Larson 1978: 190–93) as well as what were believed to have been the last remaining bison in the Big Horn Basin in northern Wyoming. A half century later, extreme drought in 1934–35 restricted the growth of grasses and, because of a lack of feed, again eliminated a large share of the range cattle. The combination of deep snow, high winds, and extreme cold that began in the fall of 1949 and lasted well into the spring of 1950 was disastrous to both domesticated and wild animals. When the snow finally melted, piles of animals that had died after drifting into fence corners and arroyos emerged. An abundance of moisture at the right times in the spring and summer of 1987 resulted in ideal range conditions for all animals. In early June, sego lilies (Calochortus nuttalli) were so thick in many areas that, using a sharp-pointed stick, one could collect a meal of lily bulbs in less than an hour. The very next year, precipitation was so scarce that the grasses managed only a few short stems; the sego lilies lay dormant until the following year, when conditions improved. In the same year as the drought, 1988, fires burned a large part of Yellowstone National Park and adjacent areas. Fortunately, the following winter of 1988–89 was not severe, and livestock and wild animals were able to survive with few losses.

If it were not for eyewitness testimony (which in the modern world becomes historic accounts), we would not be aware of these four climatic events that were short-lived yet had drastic consequences. If we could use only the methodology now employed in tree ring analysis, they would remain largely undetected: yet similar climatic events certainly occurred throughout prehistory. How they affected prehistoric hunters and hunting is far from being understood; but if they caused the same kinds of problems for them as they have historically, the effects were significant.

I find myself compelled to repeat the same thoughts with which I began this book, that too many investigators think they have analyzed hunting by declaring that prehistoric humans addressed their hunger simply by "leaving camp and killing an animal." The events leading up to the introduction of the products from hunting into the archaeological context are too often merely lumped into a category labeled *hunting* and, from that point on, largely ignored. I strongly believe this to be an unwarranted simplification: "killing an animal" hardly describes the body of learned behavior acquired over a long period of time that leads to that final act. To me, it seems that human hunting has wrongly been viewed as a kind of instinctive behavior not worthy of serious anthropological study. In reality, the volume of cultural content embedded in prehistoric hunting and its deep entanglement within many segments of institutionalized human activities largely determined the day-to-day behavior of hunting groups. And because animal behavior was a major determinant of hunting strategies, its study can be justified as a means of better understanding not only prehistoric hunting but many other aspects of prehistoric culture as well.

There are many ways to conceptualize prehistoric hunting. It can be maneuvering a bison herd into position to stampede them over a precipice, moving a few animals up the bottom of an arroyo until forward progress is terminated by a perpendicular headcut, or driving them between converging fences into a corral. It can be driving a herd of pronghorn into a circular corral made of brush with a rope stretched along its top and running them to exhaustion. It can be driving a small group of mountain sheep between converging fences that lead to a wooden ramp camouflaged with dirt and stones that leaves them no escape except a leap into a small catch pen, or driving a similar group into a net where they become immobilized long enough to be killed with clubs. It can be setting up winter quarters near a wintering herd of mule deer and killing an animal or two as needed. It can be a lone Shoshone setting traps and snares for wood rats and moving to another location as the supply of animals is temporarily exhausted. A complete list would fill many pages.

From the time the hunter leaves camp until he lays aside his weapon to field dress his kill, his movements are determined largely by the behavior of the animal involved. Even this statement is a gross oversimplification because of the wide range of interspecies behavior influenced by conditions both internal and external to the animal. Included in the former are sex, age, size, and bodily condition; among the latter are time of day, time of year, weather, terrain, vegetative cover, predators, and location of water sources. The hunter must continually evaluate these variables, along with many others, to choose the optimum strategy for obtaining the animal or animals targeted for the day's hunt. Not every attempt is successful, but over the long term, the animals have little chance of escape from an experienced hunter.

There are many possible avenues for further research into prehistoric hunters and hunting. Relatively little of the high country has had other than superficial coverage by investigators, and even a short detour from an established trail can yield new evidence. In the summer of 2002, I was made aware of a mountain sheep trap in the Absaroka Mountains in northwest Wyoming within 100 meters of a trail used by large numbers of horseback riders and backpackers during the summer and by hunters in the fall. Had an observer not deviated slightly from the trail, that trap might not have been located before it had completely deteriorated. Its location demonstrated a somewhat different strategy of handling mountain sheep and has taken us one step closer to a better understanding of prehistoric hunting. And as each new trap is studied, the locations of others are becoming more predictable.

There is a delicate balance between preservation and loss of evidence of prehistoric human hunting. Careful inspection of the sides of arroyos frequently reveals bones, some of which represent human kills. This kind of archaeological evidence is often rapidly destroyed by events such as flash floods and spring snow melt that can change the configuration of the arroyo enough to remove all signs of human activity. A lightning-caused fire can quickly erase all evidence of a structure such as a mountain sheep trap, as happened during the disastrous fires of 1988. Despite the commonalities of such prehistoric hunting sites, each one still has unique qualities and is in need of careful study.

The plains and mountains together truly provided a home for the large mammals and their prehistoric human predators, but many differences in archaeological materials recovered from the plains and those found in the mountains remain unexplained and are waiting to be resolved (Frison 1992; Davis and Greiser 1992). These are problems that cannot be solved by archaeologists alone: they will require significant multidisciplinary cooperation. It is exciting to contemplate what the next decade or so of research in the plains and mountains might reveal about prehistoric human mammal hunters.

Bibliography

Agenbroad, L. D.

1978a Buffalo jump complexes in Owyhee County, Idaho. In *Bison procurement and utilization: A symposium*, edited by L. B. Davis and M. Wilson, 213–21. Plains Anthropologist Memoir 14. Lincoln, Neb.: Augstums Printing Service.

1978b *The Hudson-Meng site: An Alberta kill in the Nebraska High Plains.* Washington, D.C.: University Press of America.

1990 The mammoth population of the Hot Springs site and associated fauna. In *Megafauna and man: Discovery of America's heartland*, edited by L. D. Agenbroad, J. I. Mead, and L. W. Nelson, 32–39. Hot Springs: Mammoth Site of Hot Springs, South Dakota.

Albanese, J.

1971 Geology of the Ruby site area, Wyoming, 48CA302. *American Antiquity* 36(1): 91–95.

1986 The geology and soils of the Colby site. In *The Colby site: Taphonomy and archaeology of a Clovis mammoth kill in northern Wyoming*, edited by G. C. Frison and L. C. Todd, 143–63. Albuquerque: University of New Mexico Press.

1987 Geological investigations. In *The Horner site: The type site of the Cody Complex*, edited by G. C. Frison and L. C. Todd, 279–326. Orlando, Fla.: Academic Press.

Anderson, E.

1968 Fauna of the Little Box Elder Cave, Converse County, Wyoming. *University of Colorado Studies, Series in Earth Sciences* 6: 1–59.

Anell, B.
1969 *Running down and driving of game in North America.* Studia Ethnographica Upsaliensia, 30. Uppsala: Inst. för allm. och jämförande etnografi.

Arkush, B. S.
1986 Aboriginal exploitation of pronghorn in the Great Basin. *Journal of Ethnobiology* 6(2): 239–55.

Beal, M. D.
1963 *"I will fight no more forever": Chief Joseph and the Nez Perce War.* Seattle: University of Washington Press.

Bement, L. C.
1999 *Bison hunting at Cooper site: Where lightning bolts drew thundering herds.* Norman: University of Oklahoma Press.

Bement, L. C., B. Carter, and K. Buehler
2002 Jake Bluff: A Clovis bison kill in northwest Oklahoma. Paper presented at the 60th Plains Anthropological Conference, Oklahoma City, Okla., October 23–26.

Benedict, J.
1996 *The game drives of Rocky Mountain National Park.* Research Report no. 7. Ward, Colo.: Center for Mountain Archaeology.

Bentzen, R. C.
1962 The Powers-Yonkee bison trap. *Plains Anthropologist* 7(18): 113–18.

Berger, J., J. E. Swenson, and I. Persson
2001 Recolonizing carnivores and naïve prey: Conservation lessons from Pleistocene extinctions. *Science* 291: 1036–39.

Bergman, C. A., and E. McEwen
1997 Sinew-reinforced and composite bows. In *Projectile Technology,* edited by H. Knecht, 143–60. New York: Plenum.

Berman, J. E.
1959 Bison bones from the Allen site, Wyoming. *American Antiquity* 25(1): 116–17.

Binford, L. R.
1978 *Nunamiut ethnoarchaeology.* New York: Academic Press.
1981 *Bones: Ancient men and modern myths.* New York: Academic Press.

Bleed, P.
1986 The optimal design of hunting weapons: Maintainability or reliability. *American Antiquity* 51(4): 737–47.

Bradley, B. A.
1991 Lithic technology. In *Prehistoric hunters of the High Plains,* edited by G. C. Frison, 369–95. 2nd ed. San Diego: Academic Press.

Bradley, B. A., and G. C. Frison
1996 Flaked-stone and worked-bone artifacts from the Mill Iron site. In *The Mill Iron site,* edited by G. C. Frison, 43–70. Albuquerque: University of New Mexico Press.

Brink, J. W., and M. Rollans
   1990 Thoughts on the structure and function of drive lane systems at communal buffalo jumps. In *Hunters of the recent past,* edited by L. B. Davis
   and B. O. K. Reeves, 152–63. London: Unwin Hyman.

Brown, B.
   1932 The buffalo drive. *Natural History* 32(1): 75–82.

Brumley, J. H.
   1984 The Laidlaw site: An aboriginal antelope trap from southeastern Alberta. In *Archaeological survey of Alberta,* edited by D. Burley, 78–96.
   Occasional Paper no. 23. [Edmonton]: Alberta Culture, Historical Resources Division.

Bryan, K., and L. L. Ray
   1940 *Geologic antiquity of the Lindenmeier site in Colorado.* Smithsonian
   Miscellaneous Collections 99, no. 2. Washington, D.C.: Smithsonian Institution.

Bryant, L. D., and C. Maser
   1982 Classification and distribution. In *Elk of North America,* edited by
   J. W. Thomas and D. E. Toweill, 1–60. Harrisburg, Pa.: Stackpole Books.

Bupp, S. L.
   1981 The Willow Springs Bison Pound: 48AB130. Master's thesis, University of Wyoming, Laramie.

Butler, B. R.
   1963 An Early Man site at Big Camas Prairie, south-central Idaho. *Tebiwa*
   6(1): 22–33.

Byers, C. R., and G. A. Bettas (eds.)
   1999 *Records of North American big game.* 11th ed. Missoula, Mont.: Boone
   and Crockett Club.

Canby, T. Y.
   1979 The search for the first Americans. *National Geographic* 156(3):
   330–63.

Chittenden, H. R., and A. T. Richardson (eds.)
   1905 *Life, letters, and travels of Father Pierre-Jean De Smet, S.J., 1801–1873.*
   Vol. 3. New York: Francis P. Harper.

Chorn, J., B. A. Frase, and C. D. Frailey
   1988 Late Pleistocene pronghorn, *Antilocapra americana,* from Natural Trap
   Cave, Wyoming. *Transactions of the Nebraska Academy of Sciences* 26:
   127–39.

Conner, S. W.
   1970 Elk antler piles made by Indians on Northwestern Plains. Manuscript
   in possession of the author.

Cosgrove, C. B.
   1947 *Caves of the Upper Gila and Hueco areas of New Mexico and Texas.*
   Papers of the Peabody Museum of American Archaeology and Ethnology,
   Harvard University, 24, no. 2. Cambridge, Mass.: The Museum.

Cotter, J. L.

1937 The occurrence of flints and extinct animals in pluvial deposits near Clovis, New Mexico, Part IV, Report on the excavations at the Gravel Pit in 1936. *Proceedings of the Philadelphia Academy of Natural Sciences* 89: 2–16.

1938 The occurrence of flints and extinct animals in pluvial deposits near Clovis, New Mexico, Part VI, Report on field season of 1937. *Proceedings of the Philadelphia Academy of Natural Sciences* 90: 113–17.

Cowie, I.

1913 *The company of adventurers: A narrative of seven years in the service of the Hudson's Bay Company during 1867–1874.* Toronto: William Briggs.

Craighead, J. J., J. S. Summer, and J. A. Mitchell

1995 *The grizzly bears of Yellowstone.* Washington, D.C.: Island Press.

Damon, P. E., C. V. Haynes, Jr., and A. Long

1964 Arizona radiocarbon dates. *Radiocarbon* 6: 91–107.

Daniel, G. E.

1952 *A hundred years of archaeology.* London: Gerald Duckworth.

Davis, L. B.

1971 The Lindsay Mammoth site (24DW501): Paleontology and paleoecology. In *Abstracts, 29th Plains Anthropological Conference,* 5–7. Winnipeg, Man., October 9–11.

1978 The 20th-century commercial mining of Northern Plains bison kills. In *Bison procurement and utilization: A symposium,* edited by L. B. Davis and M. Wilson, 254–86. Plains Anthropologist Memoir 14. Lincoln, Neb.: Augstums Printing Service.

1982 *Archaeology and geology of the Schmitt Chert Mine, Missouri headwaters: Guidebook for field trip held in conjuncton with the 35th annual meeting of Rocky Mountain Section of the Geological Society of America.* Bozeman: Department of Earth Sciences, Montana State University.

Davis, L. B., J. W. Fisher, Jr., M. C. Wilson, S. Chomko, and R. E. Morlan

2000 Avonlea Phase winter fare at Lost Terrace, upper Missouri River Valley of Montana: The vertebrate fauna. In *Pronghorn past and present: Archaeology, ethnography, and biology,* edited by J. V. Pastor and P. M. Lubinski, 53–69. Plains Anthropologist Memoir 32. Lincoln, Neb.: Plains Anthropological Society.

Davis, L. B., and S. T. Greiser

1992 Indian Creek Paleoindians: Early occupation of the Elkhorn Mountains' southeast flank, west-central Montana. In *Ice Age hunters of the Rockies,* edited by D. J. Stanford and J. S. Day, 225–84. [Denver]: Denver Museum of Natural History; Niwot: University Press of Colorado.

Davis, L. B., and E. Stallcop

1966 *The Wahkpa Chu'gn site (24HL101): Late hunters in the Milk River Valley, Montana.* Montana Archaeological Society Memoir 3. [Missoula.]

Davis, L. B., and C. D. Zeier
1978 Multi-phase Late Period bison procurement at the Antonsen site, South-western Montana. In *Bison procurement and utilization: A symposium,* edited by L. B. Davis and M. Wilson, 222–35. Plains Anthropologist Memoir 14. Lincoln, Neb.: Augstums Printing Service.

Denhardt, R. M.
1947 *The horse of the Americas.* Norman: University of Oklahoma Press.

Denig, E. T.
1930 *Indian Tribes of the Upper Missouri.* Edited by J. N. B. Hewitt. Forty-sixth annual report, U.S. Bureau of American Ethnology. Washington, D.C.

Dibble, D. S., and D. Lorrain
1968 *Bonfire Shelter: A stratified bison kill site in the Amistad Reservoir area, Val Verde County, Texas.* Texas Memorial Museum, Miscellaneous Papers 1. Austin: [University of Texas, Texas Memorial Museum].

Dixon, E. J.
2000 Coastal navigators. *Discovering Archaeology* 2(1): 34–35.

Dominick, D.
1964 The Sheepeaters. *Annals of Wyoming* 36(2): 131–68.

Driver, H. E.
1961 *Indians of North America.* Chicago: University of Chicago Press.

Driver, J. C.
1985 Prehistoric hunting strategies in the Crowsnest Pass, Alberta. *Canadian Journal of Archaeology* 6(2): 109–29.

Eakin, D. H.
1989 Report of archaeological test excavations at the Pagoda Creek Site 48PA853. Report on file at the Office of Wyoming State Archaeologist, Department of Anthropology, University of Wyoming, Laramie.

Efremov, J. A.
1940 Taphonomy, a new branch of paleontology. *Pan American Geologist* 74: 81–93.

Egan, H. R.
1917 *Pioneering the West, 1846 to 1878: Major Howard Egan's Diary.* Edited by W. M. Egan. Salt Lake City: Skelton Publishing.

Ewers, J. C.
1949 The last bison drives of Blackfoot Indians. *Journal of the Washington Academy of Sciences* 39: 355–60.

1955 *The horse in Blackfoot Indian culture, with comparative material from other western tribes.* Bureau of American Ethnology, Bulletin 159. Washington, D.C.: U.S. Government Printing Office.

Ferris, W. A.
1940 *Life in the Rocky Mountains.* Edited by P. C. Phillips. Denver: Old West Publishing Company.

Fidler, P.

n.d. *Journal of a journey over land from Buckingham House to the Rocky Mountains in 1792 and 1793*. Manuscript A.36/6 in Hudson's Bay Company Archives, London.

Figgins, J. D.

1927 The antiquity of man in America. *Natural History* 27(3): 229–39.

1933 A further contribution to the antiquity of man in America. *Proceedings of the Colorado Museum of Natural History* 12(2): 4–8.

Fisher, J. W., Jr.

1984 Medium-sized artiodactyl butchering and processing. In The Dead Indian Creek site: An Archaic occupation in the Absaroka Mountains of northwestern Wyoming, edited by G. C. Frison and D. N. Walker, 63–82. *Wyoming Archaeologist* 27(1–2): 11–122.

Fisher, J. W., Jr., and G. C. Frison

2000 Site structure and zooarchaeology at the Boar's Tusk site, Wyoming. In *Pronghorn past and present: Archaeology, ethnography, and biology*, edited by J. V. Pastor and P. M. Lubinski, 89–108. Plains Anthropologist Memoir 32. Lincoln, Neb.: Plains Anthropological Society.

Forbis, R. G.

1962a Old Women's Buffalo Jump, Alberta. *National Museum of Canada Bulletin*, no. 180: 56–123.

1962b A stratified buffalo kill in Alberta. In *Symposium on buffalo jumps*, edited by C. Malouf and S. Conner, 3–7. Montana Archaeological Society Memoir no. 1. Missoula.

1968 Fletcher: A Paleo-Indian site in Alberta. *American Antiquity* 33: 1–10.

1985 The McKean Complex as seen from Signal Butte. In *McKean/Middle Plains Archaic current research*, edited by M. Kornfeld and L. C. Todd, 21–29. Occasional Papers on Wyoming Archaeology 4. Laramie: Department of Anthropology, University of Wyoming.

Francis, J. E.

1991 An overview of Wyoming rock art. In *Prehistoric hunters of the High Plains*, edited by G. C. Frison, 397–430. 2nd ed. San Diego: Academic Press.

Francis, J. E., and L. L. Loendorf

2002 *Ancient visions*. Salt Lake City: University of Utah Press.

Fremont, General J. C.

1887 *Memoirs of my life*. Chicago: Belford, Clark.

Frison, G. C.

1962 Wedding of the Waters Cave: A stratified site in the Bighorn Basin of northern Wyoming. *Plains Anthropologist* 7(18): 246–65.

1965 Spring Creek Cave. *American Antiquity* 31(1): 81–94.

1967a Archaeological evidence of the Crow Indians in northern Wyoming: A study of a Late Prehistoric buffalo economy. Ph.D. diss., University of Michigan, Ann Arbor.

1967b *The Piney Creek sites, Wyoming (48 JO 311 and 312).* University of Wyoming Publications 33, no. 1. Laramie: University of Wyoming.

1968a Daugherty Cave, Wyoming. *Plains Anthropologist* 13(42): 253–95.

1968b Site 48SH312: An Early Middle Period bison kill in the Powder River Basin of Wyoming. *Plains Anthropologist* 13(39): 31–39.

1970a *The Glenrock Buffalo Jump, 48CO304: Late Prehistoric buffalo procurement and butchering.* Plains Anthropologist Memoir 7. [Lincoln, Neb.: Plains Anthropologist.]

1970b The Kobold Site, 24BH406: A post-Altithermal record of buffalo-jumping for the Northwestern Plains. *Plains Anthropologist* 15(47): 1–35.

1971a The buffalo pound in Northwestern Plains prehistory: Site 48CA302, Wyoming. *American Antiquity* 36(1): 77–91.

1971b Shoshonean antelope procurement in the upper Green River Basin, Wyoming. *Plains Anthropologist* 16(54): 258–84.

1973 *The Wardell buffalo trap 48SU301: Communal procurement in the upper Green River Basin, Wyoming.* Anthropological Papers, Museum of Anthropology, University of Michigan, no. 48. Ann Arbor: University of Michigan.

1974 (ed.) *The Casper site: A Hell Gap bison kill on the High Plains.* New York: Academic Press.

1978 *Prehistoric hunters of the High Plains.* New York: Academic Press.

1979 Observations on the use of stone tools: Dulling of working edges of some chipped stone tools in bison butchering. In *Lithic use-wear analysis,* edited by B. Hayden, 259–68. New York: Academic Press.

1980 A composite, reflexed, mountain sheep horn bow from western Wyoming. *Plains Anthropologist* 25(88): 173–76.

1982a Bison procurement. In *The Agate Basin site: A record of the Paleoindian occupation of the Northwestern High Plains,* edited by G. C. Frison and D. J. Stanford, 263–69. New York: Academic Press.

1982b Folsom components. In *The Agate Basin site: A record of the Paleoindian occupation of the Northwestern High Plains,* edited by G. C. Frison and D. J. Stanford, 37–76. New York: Academic Press.

1982c Introduction to *The Agate Basin site: A record of the Paleoindian occupation of the Northwestern High Plains,* edited by G. C. Frison and D. J. Stanford, 1–26. New York: Academic Press.

1982d Paleoindian winter subsistence strategies on the High Plains. In *Plains Indian studies: A collection of essays in honor of John C. Ewers and Waldo R. Wedel,* edited by D. H. Ubelaker and H. J. Viola, 193–201. Smithsonian Contribution to Anthropology, no. 30. Washington, D.C.: Smithsonian Institution Press.

1982e A probable Paleoindian flintknapping kit from the Medicine Lodge Creek site 48BH499, Wyoming. *Lithic Technology* 9(1): 3–5.

1982f The Sheaman site: A Clovis component. In *The Agate Basin Site: A record of the Paleoindian occupation of the Northwestern High Plains,* edited by G. C. Frison and D. J. Stanford, 143–56. New York: Academic Press.

1983 The Lookingbill site 48FR308. *Tebiwa* 20: 1–16.

1984 The Carter/Kerr-McGee Paleoindian site: Cultural resource management and archaeological research. *American Antiquity* 49(2): 288–314.

1987 Prehistoric, plains-mountain, large mammal, communal hunting strategies. In *The evolution of human hunting,* edited by M. Nitecki and D. Nitecki, 177–223. New York: Plenum.

1989 Experimental use of Clovis weaponry and tools on African elephants. *American Antiquity* 54(4): 766–84.

1991a Hunting strategies, prey behavior, and mortality data. In *Human predators and prey mortality,* edited by M. C. Stiner, 5–30. Boulder, Colo.: Westview Press.

1991b *Prehistoric hunters of the High Plains.* 2nd ed. San Diego: Academic Press.

1992 The foothills-mountains and the open plains: The dichotomy in Paleoindian subsistence strategies between two ecosystems. In *Ice Age hunters of the Rockies,* edited by D. J. Stanford and J. S. Day, 323–32. [Denver]: Denver Museum of Natural History; Niwot: University Press of Colorado.

1996 (ed.) *The Mill Iron site.* Albuquerque: University of New Mexico Press.

1997 *Camelops* on the Northern Plains: When did they become extinct and were they hunted by North American Paleoindians? In *Proceedings of the 1993 Bone Modification Conference, Hot Springs, South Dakota,* edited by L. A. Hannus, L. Rossum, and R. P. Winham, 12–23. Occasional Publication no. 1. Sioux Falls, S.D.: Archaeology Laboratory, Augustana College.

1998 Paleoindian large mammal hunters on the plains of North America. *Proceedings, National Academy of Sciences* 95: 14576–83.

2000a A C14 date on a Late-Pleistocene *Camelops* at the Casper–Hell Gap site. *Current Research in the Pleistocene* 17: 28–29.

2000b The Eden-Farson pronghorn kill 48SW304: Taphonomic analysis and animal behavior. In *Pronghorn past and present: Archaeology, ethnography, and biology,* edited by J. V. Pastor and P. M. Lubinski, 29–38. Plains Anthropologist Memoir 32. Lincoln, Neb.: Plains Anthropological Society.

Frison, G. C., R. L. Andrews, J. M. Adovasio, R. C. Carlisle, and R. Edgar
1986 A late Paleoindian animal trapping net from northern Wyoming. *American Antiquity* 51(2): 352–61.

Frison, G. C., and B. A. Bradley
1980 *Folsom tools and technology at the Hanson site, Wyoming.* Albuquerque: University of New Mexico Press.

1981 Fluting Folsom points: Archaeological evidence. *Lithic Technology* 10(1): 13–16.

1999 *The Fenn Cache: Clovis weaons and tools.* Santa Fe, N.M.: One Horse Land and Cattle Company.

Frison, G. C., and C. Craig
1982 Bone, antler, and ivory artifacts and manufacture technology. In *The Agate Basin site: A record of the Paleoindian occupation of the Northwestern High Plains,* edited by G. C. Frison and D. J. Stanford, 157–73. New York: Academic Press.

Frison, G. C., C. A. Reher, and D. N. Walker
1990 Prehistoric mountain sheep hunting in the Central Rocky Mountains of North America. In *Hunters of the recent past,* edited by L. B. Davis and B. O. K. Reeves, 208–40. London: Unwin Hyman.

Frison, G. C., and D. J. Stanford
1982a (eds.) *The Agate Basin site: A record of the Paleoindian occupation of the Northwestern High Plains.* New York: Academic Press.

1982b Summary and conclusions. In *The Agate Basin site: A record of the Paleoindian occupation of the Northwestern High Plains,* edited by G. C. Frison and D. J. Stanford, 361–70. New York: Academic Press.

Frison, G. C., and L. C. Todd (eds.)
1986 *The Colby mammoth site: Taphonomy and archaeology of a Clovis kill in northern Wyoming.* Albuquerque: University of New Mexico Press.

1987 *The Horner site: The type site of the Cody Cultural Complex.* Orlando: Academic Press.

Frison, G. C., and D. N. Walker (eds.)
1984 The Dead Indian Creek site: An Archaic occupation in the Absaroka Mountains of northwest Wyoming. *Wyoming Archaeologist* 27(1–2): 11–122.

Frison, G. C., M. Wilson, and D. N. Walker
1978 *The Big Goose Creek site: Bison procurement and faunal analysis.* Occasional Papers on Wyoming Archaeology no. 1. Laramie: Wyoming State Archaeologist's Office.

Frison, G. C., M. Wilson, and D. J. Wilson
1976 Fossil bison and artifacts from an Early Altithermal period arroyo trap in Wyoming. *American Antiquity* 41(1): 28–57.

Frison, G. C., and G. M. Zeimens
1980 Bone projectile points: An addition to the Folsom Cultural Complex. *American Antiquity* 45(2): 231–37.

Geist, V.
1981 Behavior: Adaptive strategies in mule deer. In *Mule and black-tailed deer of North America,* edited by O. C. Wallmo, 157–223. Lincoln: University of Nebraska Press.

1982 Adaptive behavioral strategies. In *Elk of North America,* edited by J. W. Thomas and D. E. Toweill, 219–77. Harrisburg, Pa.: Stackpole Books.

Gilbert, B. M., and L. D. Martin
1984 Late Pleistocene fossils of Natural Trap Cave, Wyoming, and the climatic model of extinction. In *Quaternary extinctions,* edited by P. S. Martin and R. G. Klein, 138–47. Tucson: University of Arizona Press.

Gilmore, M. R.
1924 Old Assiniboine buffalo-drive in North Dakota. *Indian Notes* 1: 204–11.

Gladwin, H. S.
1947 *Men out of Asia.* New York: McGraw-Hill.

Gramley, R. M.
1990 *Guide to the Palaeo-Indian artifacts of North America.* Buffalo, N.Y.: Persimmon Press.

1993 *Richey Clovis Cache: Earliest Americans along the Columbia River.* Buffalo, N.Y.: Persimmon Press.

Greaves, R. D.
1997 Hunting and multifunctional use of bows and arrows: Ethnoarchaeology of technological organization among Pume hunters of Venezuela. In *Projectile Technology,* edited by H. Knecht, 287–320. New York: Plenum Press.

Grinnell, G. B.
1923 *The Cheyenne Indians, their history and ways of life.* Vol. 1. New Haven: Yale University Press.

1961 *Pawnee, Blackfoot, and Cheyenne.* New York: Charles Scribner's Sons.

Haines, F.
1938a The northward spread of horses among the Plains Indians. *American Anthropologist* 40(3): 429–37.

1938b Where did the Plains Indians get their horses? *American Anthropologist* 40(1): 112–17.

1970 *The buffalo.* New York: Thomas Y. Crowell.

Hall, E. R.
1981 *The mammals of North America.* New York: John Wiley and Sons.

Hannus, L. A.
1990 Mammoth hunting in the New World. In *Hunters of the recent past,* edited by L. B. Davis and B. O. K. Reeves, 47–67. London: Unwin Hyman.

Harrington, M. R.
1933 *Gypsum Cave, Nevada.* Southwest Museum Papers, no. 8. Los Angeles: Southwest Museum.

Harris, A. H.
2002 The Mummy Cave tetrapods. In *The archeology of Mummy Cave, Wyoming: An introduction to Shoshonean prehistory,* edited by W. M. Husted and R. Edgar, 163–70. National Park Service, Special Report no. 4. Lincoln, Neb.: Midwest Archaeological Center.

Haspel, H., and G. C. Frison
1987 The Finley site bison bone. In *The Horner site: The type site of the Cody*

Cultural Complex, edited by G. C. Frison and L. C. Todd, 475–91. Orlando: Academic Press.

Haury, E. W., E. B. Sayles, and W. H. Wasley
1959 The Lehner mammoth site, southeastern Arizona. *American Antiquity* 25(1): 2–30.

Hawkes, K.
1990 Why do men hunt? Benefits for risky choices. In *Risk and uncertainty in tribal and peasant economies,* edited by E. Cashdan, 145–66. Boulder, Colo.: Westview Press.

Haynes, C. V., Jr.
1993 Clovis-Folsom geochronology and climatic change. In *From Kostenki to Clovis: Upper Paleolithic-Paleo-Indian adaptations,* edited by O. Soffer and N. D. Praslov, 219–36. New York: Plenum Press.

Hendry, M. H.
1983 *Indian rock art in Wyoming.* Lincoln, Neb.: Augstums Printing Service.

Henry, A., and D. Thompson
1897 *New light on the early history of the Greater Northwest.* Edited by E. Coues. 3 vols. New York: Francis P. Harper.

Hill, M. G., G. C. Frison, and D. N. Walker
1999 Folsom pronghorn utilization at Agate Basin, Wyoming. Paper presented at the symposium *Pronghorn past and present.* Western Wyoming College, Rock Springs, Wyo., September 17–18.

Hilman, R.
1984 Artifact and feature descriptions. In The Dead Indian Creek Site: An Archaic occupation in the Absaroka Mountains of northwestern Wyoming, edited by G. C. Frison and D. N. Walker, 23–50. *Wyoming Archaeologist* 27(1–2): 11–122.

Hofman, J. L.
2000 The Clovis hunters. *Discovering Archaeology* 2(1): 42–44.

Hofman, J. L., and R. W. Graham
1998 The Paleo-Indian cultures of the Great Plains. In *Archaeology of the Great Plains,* edited by W. R. Wood, 87–140. Lawrence: University Press of Kansas.

Hogan, B.
1974 Two high altitude game trap sites in Montana. Master's thesis, University of Montana, Missoula.

Holmes, W. H.
1919 *Handbook of aboriginal American antiquities.* Bureau of American Ethnology, Bulletin 60. Washington: Government Printing Office.

Honess, F., and N. Frost
1942 *Wyoming bighorn sheep study.* Wyoming Game and Fish Department, Bulletin no. 1. Cheyenne.

Howard, E. B.
1943 The Finley site: Discovery of Yuma points *in situ* near Eden, Wyoming. *American Antiquity* 8(3): 224–34.

Howard, E. B., L. Satterthwaite, Jr., and C. Bache
1941 Preliminary report on a buried Yuma site in Wyoming. *American Antiquity* 7(1): 70–74.

Huckell, B. B.
1979 Of chipped stone tools, elephants, and the Clovis hunters: An experiment. *Plains Anthropologist* 24(85): 177–89.

1982 The Denver elephant project: A report on experimentation with thrusting spears. *Plains Anthropologist* 27(97): 217–24.

Hughes, S. S.
1981 Projectile point variability: A study of point curation at a Besant kill in south central Wyoming. Master's thesis, University of Wyoming, Laramie.

2000 The Sheepeater myth of northwestern Wyoming. *Plains Anthropologist* 45(171): 63–83.

Husted, W. M., and R. Edgar
2002 *The archeology of Mummy Cave, Wyoming: An introduction to Shoshonean prehistory.* Southeast Archeological Center, Technical Reports Series, no. 9. Lincoln, Neb.: U.S. Dept. of the Interior, National Park Service, Midwest Archeological Center.

Hyde, G. E.
1974 *Spotted Tail's folk: A history of the Brule Sioux.* 2nd ed. Norman: University of Oklahoma Press.

Irwin, H. T., and G. A. Agogino
1962 Ice Age man vs. mammoth in Wyoming. *National Geographic* 121(6): 828–37.

Irwin-Williams, C., H. T. Irwin, G. Agogino, and C. V. Haynes, Jr.
1973 Hell Gap: Paleo-Indian occupation on the High Plains. *Plains Anthropologist* 18(59): 40–53.

Jameson, J. H.
1984 Artifact and feature descriptions. In The Dead Indian Creek site: An Archaic occupation in the Absaroka Mountains of northwestern Wyoming, edited by G. C. Frison and D. N. Walker, 23–50. *Wyoming Archaeologist* 27(1–2): 11–122.

Jepsen, G. L.
1953 Ancient buffalo hunters of northwestern Wyoming. *Southwestern Lore* 19: 19–25.

Jodry, M. A., and D. J. Stanford
1992 Stewart's Cattle Guard site: An analysis of bison remains in a Folsom kill-butchery campsite. In *Ice Age hunters of the Rockies,* edited by D. J. Stanford and J. S. Day, 101–68. [Denver]: Denver Museum of Natural History; Niwot: University Press of Colorado, Boulder.

Johnson, E.

1987 Cultural activities and interactions. In *Late Quaternary studies on the Southern High Plains,* edited by E. Johnson, 120–58. College Station: Texas A&M University Press.

1989 Human modified bone from Early Southern Plains sites. In *Bone modification,* edited by R. Bonnichsen and M. Sorg, 431–72. Orono: Center for the Study of the First Americans, Institute for Quaternary Studies, University of Maine.

Kehoe, T. F.

1967 *The Boarding School Bison Drive site.* Plains Anthropologist Memoir 4. Lincoln, Neb.: Plains Anthropologist.

Keyser, J. D.

1974 The LaMarche game trap: An early historic game trap in southwestern Montana. *Plains Anthropologist* 19(55): 173–79.

Knecht, H.

1997 Projectile points of bone, antler, and stone: Experimental explorations of manufacture and use. In *Projectile technology,* edited by H. Knecht. 191–212. New York: Plenum Press.

Knight, D. H.

1994 *Mountains and plains: The ecology of Wyoming landscapes.* New Haven: Yale University Press.

Kolm, K. E.

1974 ERTS MSS imagery applied to mapping of sand dunes in Wyoming. In *Applied geology and archaeology: The Holocene history of Wyoming,* edited by M. Wilson, 34–39. Geological Survey of Wyoming, Report of Investigations, no. 10. Laramie: Geological Survey of Wyoming.

Kooyman, B., P. McNeil, L. V. Hills, and S. Tolman

2002 Stress in Late Pleistocene Northern Plains large mammal populations as seen through the Wally's Beach site, Alberta, Canada. Paper presented at the 67th annual meeting of the Society for American Archaeology, Denver, Colo., March 20–24.

Kooyman, B., M.E. Newman, C. Cluny, M. Lobb, S. Tolman, P. McNeil, and L.V. Hill

2001 Identification of horse exploitation by Clovis hunters based on protein analysis. *American Antiquity* 66(4): 686–91.

Kornfeld, M., G. C. Frison, M. L. Larson, J. C. Miller, and J. Saysette

1999 Paleoindian bison procurement and paleoenvironments in Middle Park of Colorado. *Geoarchaeology* 14: 655–74.

Kornfeld, M., M. L. Larson, D. J. Rapson, and G. C. Frison

2001 Ten thousand years in the Middle Rocky Mountains. *Journal of Field Archaeology* 28 (384): 307–24.

Kroeber, A. L.

1939 *Cultural and natural areas of North America.* Berkeley: University of California Press.

Lahren, L. A., and R. Bonnichsen
1974 Bone foreshafts from a Clovis burial in southwest Montana. *Science* 186: 147–50.

Lanier, J. L.
2001 Handling bison. In *Bison are back—2000: Proceedings of the Second International Bison Conference, August 2–4, 2000, Edmonton, AB,* edited by R. Rutley, 265–68. Leduc, Alta.: Bison Centre of Excellence.

Larson, T. A.
1978 *History of Wyoming.* 2nd ed. Lincoln: University of Nebraska Press.

Laubin, R., and G. Laubin
1980 *American Indian archery.* Civilization of the American Indian Series, 154. Norman: University of Oklahoma Press.

Laughlin, W. S.
1968 Hunting: An integrating biobehavior system and its evolutionary importance. In *Man the hunter,* edited by R. B. Lee and I. DeVore, 304–20. Chicago: Aldine.

Laws, R. M., I. S. C. Parker, and R. C. B. Johnstone
1975 *Elephants and their habitats: The ecology of elephants in north Bunyoro, Uganda.* Oxford: Clarendon Press.

Lee R. B., and I. DeVore (eds.)
1968 *Man the hunter.* Chicago: Aldine.

Leonhardy, F. C.
1966 *Domebo: A Paleo-Indian mammoth kill in the Prairie Plains.* Contributions of the Museum of the Great Plains, no. 1. [Lawton, Okla.: Great Plains Historical Association].

Lewis, H. P.
1947 *Buffalo kills in Montana.* Report on file at National Park Service, Midwest Archaeological Center, Lincoln, Neb.

Lewis, M.
[1904–5] 1969 *Original journals of the Lewis and Clark expedition, 1804–1806.* Edited by R. G. Thwaites. 8 vols. Reprint, New York: Arno Press.

Lippincott, K., M. Adair, D. R. Byrne, J. Theler, and R. E. Warren
1996 *A Late Prehistoric period pronghorn hunting camp in the southern Black Hills, South Dakota: Site 39FA23.* Special Publication of the South Dakota Archaeological Society, no. 11. Sioux Falls: South Dakota Archaeological Society.

Lippincott, K., and D. R. Byrne
1996 Vertebrate faunal remains. In *A Late Prehistoric period pronghorn hunting camp in the southern Black Hills, South Dakota: Site 39FA23,* by K. Lippincott, M. Adair, et al., 66–77. Special Publication of the South Dakota Archaeological Society, no. 11. Sioux Falls: South Dakota Archaeological Society.

Lobdell, J. E.
 1973 The Scoggin site: An Early Middle Period bison kill. Master's thesis, University of Wyoming, Laramie.

Lowie, R. H.
 1909 *The Northern Shoshone.* Anthropological Papers, American Museum of Natural History 2, pt. 2. New York: American Museum of Natural History.
 1924 *Minor ceremonies of the Crow Indians.* Anthropological Papers, American Museum of Natural History 21, pt. 5. New York: American Museum of Natural History.

Lubinski, P. M.
 1997 Pronghorn intensification in the Wyoming Basin: A study of mortality patterns and prehistoric hunting strategies. Ph.D. diss., University of Wisconsin, Madison.

Lyman, R. L.
 1982 Archaeofaunas and subsistence studies. *Advances in Archaeological Method and Theory* 5: 331–93.

Malouf, C.
 1962a Notes on the Logan Buffalo Jump. In *Symposium on buffalo jumps,* edited by C. Malouf and S. Conner, 8–11. Montana Archaeological Society Memoir no. 1. Missoula.
 1962b Panel discussion on buffalo jumps. In *Symposium on buffalo jumps,* edited by C. Malouf and S. Conner, 40–56. Montana Archaeological Society Memoir no. 1. Missoula.

Malouf, C., and S. Conner (eds.)
 1962 *Symposium on buffalo jumps.* Montana Archaeological Society Memoir no. 1. Missoula.

Mandelbaum, D. G.
 1940 The Plains Cree. *Anthropological Papers, American Museum of Natural History* 37: 155–316.

Mann, C. J.
 1968 Geology of archaeological site 48SH312, Wyoming. *Plains Anthropologist* 13(39): 40–45.

Marshall, J. (dir.)
 1957 *The hunters.* Produced by the Film Study Center of Peabody Museum, Harvard University, Cambridge, Mass. 16mm, 73 minutes.

Marshall, L. G.
 1984 Who killed cock robin? An investigation of the extinction theory. In *Quaternary extinctions,* edited by P. S. Martin and R. G. Klein, 785–806. Tucson: University of Arizona Press.

Martin, P. S.
 1967 Pleistocene overkill. *Natural History* 76(10): 32–38.

Martin, P. S., and R. G. Klein (eds.)
 1984 *Quaternary extinctions.* Tucson: University of Arizona Press.

Martin, P. S., and H. E. Wright, Jr.
1967 *Pleistocene extinctions.* New Haven: Yale University Press.

Mason, O. T.
1893 North American bows, arrows, and quivers. *Smithsonian Institution, Annual Report,* 631–79. Washington, D.C.

Maycock, W. P.
1980 *Shoot 'em again.* Cody, Wyo.: Rustler Printing and Publishing.

McCabe, R. E.
1982 Elk and Indians: Historical values and perspectives. In *Elk of North America,* edited by J. W. Thomas and D. E. Toweill, 61–123. Harrisburg, Pa.: Stackpole Books.

McDonald, J. N.
1981 *North American bison: Their classification and evolution.* Berkeley: University of California Press.

McHugh, T.
1972 *The time of the buffalo.* New York: Alfred A. Knopf.

McNamee, T.
1984 *The grizzly bear.* New York: Alfred A. Knopf.

Miller, M. E.
1976 Communal bison procurement during the Middle Plains Archaic. Master's thesis, University of Wyoming, Laramie.

Miller, M. E., and G. R. Burgett
2000 The Cache Hill site (48CA61): A bison kill-butchery site in the Powder River Basin, Wyoming. *Wyoming Archaeologist* 44(1):27–43.

Miller, M. E., P. H. Sanders, and J. E. Francis (eds.)
1999 *The Trappers Point site (48SU1006): Early Archaic adaptations in the upper Green River Basin, Wyoming.* 2 vols. Cultural Resource Series (Office of the Wyoming State Archaeologist), no. 1. Laramie: Office of the Wyoming State Archaeologist.

Moss, J. H., K. Bryan, G. W. Holmes, L. Satterthwaite, Jr., H. P. Hansen, C. B. Schultz, and W. D. Frankforter
1951 *Early Man in the Eden Valley.* Museum Monographs no. 6. Philadelphia: University Museum, University of Pennsylvania.

Mulloy, W. T.
1958 *A preliminary historical outline for the Northwestern Plains.* University of Wyoming Publications 22, nos. 1–2. Laramie: Graduate School, University of Wyoming

1959 The James Allen site near Laramie, Wyoming. *American Antiquity* 25(1): 112–16.

Murdock, G. P.
1960 *Ethnographic bibliography of North America.* 3rd ed. New Haven: Human Relations Area Files.

Nelson, M. C.
1997 Projectile points: Form, function, and design. In *Projectile technology,* edited by H. Knecht, 371–84. New York: Plenum Press.

Nelson, N. C.
1942 Camping on ancient trails. *Natural History* 49: 262–67.

Nimmo, B. W.
1971 Population dynamics of a Wyoming pronghorn cohort from Eden-Farson site, 48SW304. *Plains Anthropologist* 16(54): 285–88.

Olson, E. C.
1958 Report on the fauna of Pictograph Cave. Appendix A in *A preliminary historical outline for the Northwestern Plains,* by W. T. Mulloy, 224–25. University of Wyoming Publications 32, nos. 1–2. Laramie, Graduate School, University of Wyoming.

Pastor, J. V., and P. M. Lubinski (eds.)
2000 *Pronghorn past and present: Archaeology, ethnography, and biology.* Plains Anthropologist Memoir 32. Lincoln, Neb.: Plains Anthropological Society.

Petrie, H.
1934–35 Cattle purchases from drought areas; June 1934–February 1935. A report to G. B.Thorne, Director, Division of Livestock and Feed Grains, Agricultural Adjustment Administration. Archives of American Heritage Center, University of Wyoming, Laramie.

Popowski, B., and W. E. Pyle
1982 *The hunter's book of the pronghorn antelope.* Tulsa, Okla.: Winchester Press.

Potter, D. R.
1982 Recreational use of elk. In *Elk of North America,* edited by J. W. Thomas and D. E. Toweill, 509–59. Harrisburg, Pa.: Stackpole Books.

Rapson, D. J.
1990 Pattern and process in intrasite spatial analysis: Site structural and faunal research at the Bugas-Holding site. Ph.D. diss., University of New Mexico, Albuquerque.

Rapson, D. J., and L. B. Niven
2002 Evaluating food management strategies at the Hell Gap site, Locality 1: Inferences from an incomplete faunal record. Paper presented at the 67th Annual Meeting of the Society for American Archaeology, Denver, March 20–24.

Reagan, A. B.
1934 Some notes on the history of the Uintah Basin in northeastern Utah to 1950. *Proceedings, Utah Academy of Sciences, Arts, and Letters* 11: 55–64.

Reeves, B. O. K.
1978a Bison killing in the southwestern Alberta Rockies. In *Bison procurement and utilization: A symposium,* edited by L. B. Davis and M. Wilson,

63–78. Plains Anthropologist Memoir 14. Lincoln, Neb.: Augstums Printing Service.

1978b Head-Smashed-In: 5500 years of bison jumping in the Alberta plains. In *Bison procurement and utilization: A symposium,* edited by L. B. Davis and M. Wilson, 151–74. Plains Anthropologist Memoir 14. Lincoln, Neb.: Augstums Printing Service.

Reher, C. A.

1974 Population study of the Casper site bison. In *The Casper site,* edited by G. C. Frison, 113–24. New York: Academic Press.

Reher, C. A., and G. C. Frison

1980 *The Vore site, 48CK302, a stratified buffalo jump in the Wyoming Black Hills.* Plains Anthropologist Memoir 16. [Lincoln, Neb.: Plains Anthropologist.]

Renaud, E. B.

1932 Yuma and Folsom artifacts, new material. *Proceedings, Colorado Museum of Natural History* 11(2): 5–22.

Robbins, R. L., D. E. Redfearn, and C. P. Stone

1982 Refuges and Elk Management. In *Elk of North America,* edited by J. W. Thomas and D. E. Toweill, 479–507. Harrisburg, Pa.: Stackpole Books.

Roberts, F. H. H.

1935 A Folsom Complex: A preliminary report on investigations at the Lindenmeier site in northern Colorado. *Smithsonian Miscellaneous Collections* 94: 1–35.

1936 Additional information on the Folsom Complex. *Smithsonian Miscellaneous Collections* 95: 1–38

1943 A new site. *American Antiquity* 8(3): 300.

1961 The Agate Basin Complex. In *Homenaje a Pablo Martínez del Río en el vigésimoquinto aniversario de la primera edición de "Los orígenes americanos,"* 125–32. Mexico City: Instituto Nacional de Anthropologia.

Russell, O.

1921 *Journal of a trapper.* Boise, Idaho: Syms-York.

Sahlins, M.

1972 *Stone Age economics.* Chicago: Aldine.

Sawyer, H., and F. Lindzey

2000 *Jackson Hole pronghorn study: Final report.* Laramie: Wyoming Cooperative Fish and Wildlife Research Unit.

Schullinger, J. N.

1951 Geology and chronology of the Horner site, Park County, Wyoming. Senior thesis, Department of Geology, Princeton University, Princeton, N.J.

Secoy, F. R.

1953 *Changing military patterns on the Great Plains.* American Ethnological Society, Monograph no. 21. Seattle: University of Washington Press.

Sellards, E. H.
1938 Artifacts associated with fossil elephant. *Geological Society of America, Bulletin* 49: 999–1010.

Sellards, E. H., G. L. Evans, and G. E. Meade
1947 Fossil bison and associated artifacts from Plainview, Texas. *Geological Society of America, Bulletin* 58: 927–54.

Semenov, S. A.
1964 *Prehistoric technology: An experimental study of the oldest tools and artefacts from traces of manufacture and wear.* Translated by M. W. Thompson. London: Cory, Adams, and Mackay.

Shipman, P.
1981 *Life history of a fossil.* Cambridge, Mass.: Harvard University Press.

Simpson, J. H.
1876 *Report of explorations across the Great Basin of the Territory of Utah.* Washington, D.C.: Government Printing Office.

Simpson, T.
1984 Population dynamics of mule deer. In The Dead Indian Creek site: An Archaic occupation in the Absaroka Mountains of northwestern Wyoming, edited by G. C. Frison and D. N. Walker, 83–96. *Wyoming Archaeologist* 27(1–2): 11–122.

Smith, C. S., and L. M. McNees
2000 Pronghorn and bison procurement during the Uinta Phase in southwest Wyoming. In *Pronghorn past and present: Archaeology, ethnography, and biology,* edited by J. V. Pastor and P. M. Lubinski, 71–88. Plains Anthropologist Memoir 32. Lincoln, Neb.: Plains Anthropological Society.

Spencer, R. F.
1959 *The North Alaskan Eskimo: A study in ecology and society.* Bureau of American Ethnology, Bulletin 171. Washington, D.C.: U.S. Government Printing Office.

Stafford, M. D., G. C. Frison, D. J. Stanford, and G. M. Zeimens
2003 Digging for the color of life: Paleoindian red ochre mining at the Powars II Site, Platte County, Wyoming. *Geoarchaeology* 18(1): 71–90.

Stands in Timber, J., and M. Liberty
1967 *Cheyenne memories.* Lincoln: University of Nebraska Press.

Stanford, D. J.
1978 The Jones-Miller site: An example of a Hell Gap bison procurement strategy. In *Bison procurement and utilization: A symposium,* edited by L. B. Davis and M. Wilson, 90–97. Plains Anthropologist Memoir 14. Lincoln, Neb.: Augstums Printing Service.

1979a Bison kill by Ice Age hunters. *National Geographic* 155(1): 114–19.

1979b Carving up a "mammoth" Stone Age style. *National Geographic* 155(1): 120–21.

Stanford, D. J., and B. A. Bradley
2000 The Solutrean solution. *Discovering Archaeology* 2(1): 54–55.

Steward, J. H.
1938 *Basin-Plateau aboriginal sociopolitical groups.* Smithsonian Institution, Bureau of American Ethnology, Bulletin 120. Washington, D.C.: U.S. Government Printing Office.

1941 *Culture element distributions: XIII, Nevada Shoshoni.* Anthropological Records 4, no. 2. Berkeley: University of California Press.

Storer, T. I., and L. P. Tevis, Jr.
1978 *California grizzly.* Lincoln: University of Nebraska Press.

Straus, L. G.
2000 Solutrean settlement of North America? A review of reality. *American Antiquity* 65(1): 219–26.

Stuart, R.
1935 *The discovery of the Oregon Trail: Robert Stuart's narratives of his overland trip eastward from Astoria in 1812–13.* Edited by P. A. Rollins. New York: C. Scribner's Sons.

Sundstrom, L.
1984 *Rock art of western South Dakota and the southern Black Hills.* Special publication of the South Dakota Archaeological Society, no. 9. [Vermillion]: South Dakota Archaeological Society.

2000 Cheyenne pronghorn procurement and ceremony. In *Pronghorn past and present: Archaeology, ethnography, and biology,* edited by J. V. Pastor and P. M. Lubinski, 119–32. Plains Anthropologist Memoir 32. Lincoln, Neb.: Plains Anthropological Society.

Tankersley, K. B., K. O. Tankersley, N. R. Shaffer, M. D. Hess, J. S. Benz, F. R. Turner, M. D. Stafford, G. M. Zeimens, and G. C. Frison
1995 They have a rock that bleeds: Sunrise red ochre and its early Paleoindian occurrence at the Hell Gap site, Wyoming. *Plains Anthropologist* 40(152): 185–94.

Thomas, D. H.
1983a *The archaeology of Monitor Valley.* Vol. 1, *Epistemology.* Anthropological Papers, American Museum of Natural History, 58, pt. I. New York: American Museum of Natural History.

1983b *The archaeology of Monitor Valley.* Vol. 2, *Gatecliff Shelter.* Anthropological Papers, American Museum of Natural History, 59, pt. 1. New York: American Museum of Natural History.

1988 *The archaeology of Monitor Valley.* Vol. 3, *Survey and additional excavations.* Anthropological Papers, American Museum of Natural History, 66, pt. 2. New York: American Museum of Natural History.

Thomas, D. H., and D. Mayer
1983 Behavioral faunal analysis of selected horizons. In *The archaeology of Monitor Valley,* vol. 2, *Gatecliff Shelter,* by D. Thomas, 353–91. Anthro-

pological Papers, American Museum of Natural History, 59, pt. 1. New York: American Museum of Natural History.

Thompson, D.
1962 *David Thompson's Narrative, 1784–1812.* A new edition with added material. Edited by R. Glover. Publications of the Champlain Society, 40. Toronto: Champlain Society.

Thorne, T., G. Butler, T. Varcalli, K. Becker, and S. Hayden-Wing
1979 *The status, mortality, and response to management of the bighorn sheep of Whiskey Mountain.* Wildlife Technical Report, no. 7. Cheyenne: Wyoming Game and Fish Department.

Thorne, T., N. Kingston, W. Jolley, and R. Bergstrom
1982 *Diseases of wildlife in Wyoming.* 2nd ed. Cheyenne: Wyoming Game and Fish Department.

Titmus, G.
2002 A passion for ancient technology. *Mammoth Trumpet* 17(2): 4–9.

Todd, L. C.
1987 Taphonomy of the Horner II bone bed. In *The Horner site: The type site of the Cody Cultural Complex,* edited by G. C. Frison and L. C. Todd, 107–98. Orlando, Fla.: Academic Press.

Todd, L. C., and G. C. Frison
1986 Taphonomic study of the Colby site mammoth bones. In *The Colby mammoth site: Taphonomy and archaeology of a Clovis kill in northern Wyoming,* edited by G. C. Frison and L. C. Todd, 27–90. Albuquerque: University of New Mexico Press.

Todd, L. C., D. J. Rapson, and J. L. Hofman
1996 Dentition studies of the Mill Iron and other early Paleoindian bonebed sites. In *The Mill Iron site,* edited by G. C. Frison, 145–76. Albuquerque: University of New Mexico Press.

Trenholm, V. C.
1970 *The Arapahoes, our People.* Norman: University of Oklahoma Press.

Voorhies, M. R.
1969 *Taphonomy and population dynamics of an early Pliocene vertebrate fauna, Knox County, Nebraska.* Contributions to Geology, Special Paper, no. 1. Laramie: University of Wyoming.

Walker, D. N.
1975 A cultural and ecological analysis of the vertebrate fauna from the Medicine Lodge Creek site (48BH499). Master's thesis, University of Wyoming, Laramie.

1980 The Vore Site local fauna. Appendix 1 of *The Vore site 48CK302, a stratified buffalo jump in the Wyoming Black Hill,* by C. A. Reher and G. Frison, 154–67. Plains Anthropologist Memoir 16. Lincoln, Nebraska. [Lincoln, Neb.: Plains Anthropologist.]

1982 Early Holocene vertebrate fauna. In *The Agate Basin site: A record of the Paleoindian occupation of the northwestern High Plains,* edited by G. C. Frison and D. J. Stanford, 274–308. New York: Academic Press.

1987 Horner site local fauna: Vertebrates. In *The Horner site: The type site of the Cody Cultural Complex,* edited by G. C. Frison and L. C. Todd, 327–45. Orlando, Fla.: Academic Press.

2000 Pleistocene and Holocene records of *Antilocapra americana:* A review of the FAUNMAP data. In *Pronghorn past and present: Archaeology, ethnography, and biology,* edited by J. V. Pastor and P. M. Lubinski, 13–28. Plains Anthropologist Memoir 32. Lincoln, Neb.: Plains Anthropological Society.

Walker, D. N., and G. C. Frison
1980 The late Pleistocene mammalian fauna from the Colby Mammoth Kill site. *Contributions to Geology, University of Wyoming* 19(1): 69–79.

1982 Studies on Amerindian dogs, 3: Prehistoric wolf/dog hybrids from the Northwestern Plains. *Journal of Archaeological Science* 9: 125–72.

Wallmo, O. C.
1981 Mule and black-tailed deer distribution and habitats. In *Mule and black-tailed deer of North America,* edited by O. C. Wallmo, 1–25. Lincoln: University of Nebraska Press.

Wang, X.
1982 Late Pleistocene bighorn sheep *(Ovis canadensis)* of Natural Trap Cave, Wyoming. Master's thesis, University of Kansas, Lawrence.

Weaver, K. F.
1985 The search for our ancestors. *National Geographic* 168(5): 560–623.

Webb, S. B., M. Baker, F. K. Barbour, A. Aly, and A. C. Gilbert
1952 *Records of North American big game: A book of the Boone and Crockett Club.* New York: C. Scribner's Sons.

Wedel, W. R.
1961 *Prehistoric man on the Great Plains.* Norman: University of Oklahoma Press.

Weisel, G. F.
1951 The ram's horn tree and other medicine trees of the Flathead Indians. *Montana Magazine of History* 1(3): 5–14.

Wettlaufer, B.
1955 *The Mortlach site in the Besant Valley of central Saskatchewan.* Anthropological Series, no. 1. Regina: Department of Natural Resources.

Wheat, J. B.
1967 A Paleoindian bison kill. *Scientific American* 216: 44–52.

1972 *The Olsen-Chubbuck site: A Paleo-Indian bison kill.* Memoirs of the Society for American Archaeology, no. 26. [Washington, D.C.: Society for American Archaeology.]

1979 *The Jurgens site.* Plains Anthropologist Memoir 15. [Lincoln, Neb.:] Plains Anthropologist.

Wheeler, R. P.

1995 *Archeological investigations in three reservoir areas in South Dakota and Wyoming. Part 1, Angostura Reservoir.* Reprints in Anthropology 46. Lincoln, Neb.: J&L Reprint Company.

White, T.

1953a A method of calculating the dietary percentage of various food animals utilized by aboriginal peoples. *American Antiquity* 18(4): 396–98.

1953b Observations on the butchering techniques of some aboriginal peoples: No. 2. *American Antiquity* 19(2): 160–64.

Wied-Neuwied, Maximilian, prinz von

1906 *Travels in the interior of North America.* Translated by H. E. Lloyd. Early Western Travels, 22–24. Edited by R. G. Thwaites. Cleveland: Arthur H. Clark.

Wilmsen, E. N.

1974 *Lindenmeier: A Pleistocene hunting society.* New York: Harper and Row.

Wilmsen, E. N., and F. H. H. Roberts, Jr.

1978 *Lindenmeier, 1934–1974.* Smithsonian Contributions to Anthropology, no. 24. Washington, D.C.: Smithsonian Institution Press.

Wilson, E. N., and H. F. Driggs

1919 *The White Indian boy: The story of Uncle Nick among the Shoshones.* Yonkers-on-Hudson, N.Y.: World Book.

Wilson, G. L.

1924 *The horse and the dog in Hidatsa culture.* Anthropological Papers, American Museum of Natural History 15, pt. 2. New York: American Museum of Natural History.

Wilson, M. C.

1975 Holocene fossil bison from Wyoming and adjacent areas. Master's thesis, University of Wyoming, Laramie.

Wissler, C.

1907 Diffusion of culture in the Plains of North America. In *Proceedings, Fifteenth International Congress of Americanists,* 2:39–52. Quebec.

1910 *Material culture of the Blackfoot Indians.* Anthropological Papers, American Museum of Natural History 5, pt. 1. New York: American Museum of Natural History.

1914 The influence of the horse in the development of Plains culture. *American Anthropologist,* n.s., 16(1): 1–25.

Wormington, H. M.

1957 *Ancient man in North America.* Denver Museum of Natural History, Popular Series no. 4. 4th ed. Denver, Colo.: Denver Museum of Natural History.

Wormington, H. M., and D. Ellis (eds.)

1967 *Pleistocene studies in southern Nevada.* Nevada State Museum Anthropological Papers, no. 13. Carson City: Nevada State Museum.

Wormington, H. M., and R. G. Forbis
    1965 *An introduction to the archaeology of Alberta, Canada.* Proceedings,
    no. 11. Denver, Colo.: Denver Museum of Natural History.

Wyckoff, D. G., and W. W. Dalquest
    1997 From whence they came: The paleontology of Southern Plains bison. In
    Southern Plains bison procurement and utilization from Paleoindian to His-
    toric, edited by L. C. Bement and K. Buehler, 5–32. Plains Anthropologist
    Memoir 29. Lincoln, Neb.: Plains Anthropological Society.

Zeimens, G. M.
    2001 Preliminary investigations of the Jewett Mammoth site, Wyoming. In
    Proceedings of the International Conference on Mammoth Site Studies,
    edited by D. West, 105–7. Publications in Anthropology 22. Lawrence: Uni-
    versity of Kansas, Department of Anthropology.

Index

Absaroka Mountains: mountain sheep in, 146, 152, 157, 160; mountain sheep traps in, 149, 163–64, 229; mule deer remains in, 173 *map*, 174–75; pronghorn in, 128. *See also* Sunlight Basin (Wyo.)

Agate Basin site (Wyo.): bison and pronghorn remains at, 64, 72–73, 141, 141 *figure*, 217; elk antlers from, 109, 181, 182 *figure*, 214, 216; meat cache at, 61; projectile point from, 110; significance of, 197. *See also* Sheaman Clovis site (Wyo.)

Agricultural Adjustment Administration, 116–17

Alberta (Canada): bison kill sites in, 34–35, 80, 98, 99; Clovis points and horses linked in, 50; pronghorn trap in, 140. *See also* Head-Smashed-In site (Alberta)

Alberta projectile point, 108 *figure*, 110–11

Alces sp., 18. *See also* moose

Altithermal period, 64, 87–88

American Association for the Advancement of Science, 32

Anderson, Roy, 45

Anell, Bengt, 177–78

animal kill sites, alternate uses of, 113–14. *See also* bison kill sites; corrals and traps; mountain sheep traps; *specific animals*

animal spirits. *See* ritual activities

anthropology, limits of, 32–34, 228. *See also* humans; hunters

Antilocapra americana, 39, 121. *See also* pronghorn

antlers, deer: as atlatl hook, 109; as hunter's trophy, 170; stone-flaking hammer from, 174

antlers, elk: as atlatl hook, 109; bows from, 206, 207 *figure*, 208; as clubs, 161; as digging tools, 182, 183 *figure*; items made from, 216 *figure*; piles of, 27, 183–84; present-day collection and use of, 184–86; as projectile points, 214, 216–17; wire entangled in, 180

Antonsen Buffalo Jump (Mont.), 82

Anzick Clovis cache, 214, 215 *figure*, 219

Arapaho Indians, 183–84

archaeology: evidence on hunters from, 38–41, 42 *map*, 43; geology combined with, 70, 79, 96; interests in, 29–32; interpreting record of, 37; new methods needed for, 34–35; social organization and evidence from, 226–27

Arctodus simus, 40. *See also* bear, short-faced

Arkush, Brooke, 132

arroyo bison traps: advantages of, 75; butchering at, 112; examples of, 44–48, 70–76; excavation of, 34, 70 *figure*; geological processes at, 69–70, 74; modern analogue to,

Compositor:	Integrated Composition Systems
Indexer:	Margie Towery
Illustrator:	Bill Nelson
Text:	10/13 Sabon
Display:	Sabon
Printer and binder:	Thomson-Shore, Inc.